Laboratory Manual for
General Microbiology

Third Edition
Revised Printing

Cynthia Littlejohn
University of Southern Mississippi

Kendall Hunt
publishing company

Illustrations are courtesy of Cynthia Littlejohn unless otherwise noted

Exercise 17 contributed by: Dhiraj Acharya

Images on pages 3, 8, 14, 21, 23, 24, 39, 48, 70, 105–110
are from **Microbiology Lab Manual,** 2/E by Elizabeth Carrington, Ph.D.
Copyright © 2006, 2009 by Elizabeth Carrington, Ph.D. Reprinted by
permission of Kendall Hunt Publishing Company.

Cover image © Shutterstock, Inc.

www.kendallhunt.com
Send all inquiries to:
4050 Westmark Drive
Dubuque, IA 52004-1840

Copyright © 2010, 2015, 2017 by Kendall Hunt Publishing Company

ISBN 978-1-5249-7755-9

Published in the United States of America

Contents

Microbiological Laboratory Safety Guidelines

The rules listed below are designed to protect you and the other students and instructors while in the microbiology laboratory. You will be working with live microorganisms, most of which are harmless, but others, if ingested or allowed to enter a cut or abrasion, may result in physical discomfort or a change in your state of health. These organisms will be handled in pure cultures, which means they will be present in large numbers. Even organisms that seem rather innocuous in the everyday environment may become dangerous when grown in such high numbers. Individual susceptibility to the organisms used in the lab varies, but you can protect yourself and your family by following the simple instructions listed below.

1. Wash your hands before you begin the lab exercise and just before you leave the lab.
2. Disinfect your work bench using a 10% bleach solution before you begin and after you complete the lab exercise.
3. Keep your work space neat, clean, and uncluttered. To avoid contamination, store clothing, books, purses, cell phones, and all other personal property in the cabinet under your work bench. Any contaminated material, including cell phones, will have to be autoclaved.
4. Do not eat (including gum, hard candy, and cough drops), drink, or apply make-up (including lip balm) while in the lab.
5. Pencils, pens, or any other material should never be placed in the mouth.
6. Always wear closed toe shoes and a lab coat or apron. Do not wear the lab coat or apron outside the lab.
7. Be careful of loose clothing and dangling jewelry. If you have long hair, tie it back out of the way. Loose clothing, long hair, and dangling jewelry are particularly dangerous when working with the open flame of a Bunsen burner.
8. Do not use butane lighters to ignite the Bunsen burners. Use only matches or the strikers. Constantly be aware of the placement of your Bunsen burner and turn it off immediately after you finish using it. Make sure the gas valve is in the closed position.
9. Place contaminated pipettes, pipette tips, slides, applicator sticks, and cotton swabs into designated containers of disinfectant and never put them in the trash can.
10. Place used culture tubes and plates in a designated place indicated by your instructor. Do not leave them on your work bench. These cultures need to be sterilized in the autoclave before they are discarded so you should remove all label tape from the tubes.
11. Do not throw broken glass into the trash can. There is a broken glass discard box for these items.
12. If you spill something or break something, report it immediately to your instructor. Spilled organisms should be covered with paper towels saturated in disinfectant. The towels should remain over the spill for 15 minutes; then the entire area should be cleaned with disinfectant.
13. Be careful when inserting glass pipettes into the pump. Severe cuts may occur if you try to force the pipette into the pump. It can slip and gash your hand.
14. Alert your instructor immediately if you get cut, burned, or if you contaminate any part of your body with an organism or chemical.
15. Never remove anything from the laboratory without the permission of the instructor.
16. All inoculated materials must be properly labeled with your name, the date, and your section.
17. When flaming your inoculating loop, the loop should always be placed into the base of the flame and allowed to remain until it is red hot to remove all contaminates.
18. Do not mouth pipette any substance. Do not open cultures to smell them.
19. Learn the location of the fire extinguisher, safety shower, eye wash station, and first aid kit.
20. For all serious emergencies, call 911 and you will be connected to campus police.

Background

Compound light microscopes contain a series of lenses that allow for the magnification of a specimen up to one thousand times. Magnification is limited by the resolving power of the lenses. Resolution indicates the clarity a microscope is able to achieve. This is relative to how far apart two objects must be in order to be seen as distinct or separate from each other. For most light microscopes, the limit of resolution is 0.2 μm. The use of immersion oil will reduce refraction or bending of light and increase resolution.

Light is passed through a condenser lens, which concentrates the light and focuses it on the specimen. The diaphragm regulates the amount of light passing through the condenser to the specimen. The light then passes through the objective lens, where it refracts, and creates a magnified image. The image is further magnified by the ocular lens (eyepiece). Each objective lens and ocular lens will be labeled with the magnification they are able to produce. As magnification increases the working distance (distance between the specimen and objective), your field of view (the amount of the specimen you can visualize) and the intensity of light all decrease. Therefore, small adjustments may be required to the fine focus and the condenser in order to sharpen the image.

In order to determine the total magnification of a specimen one would multiply the magnification of the objective lens by the magnification of the ocular lens. For example, if the objective lens used is capable of 100X magnification and the ocular lens is capable of 10X magnification then (100X) X (10X) = a total magnification of 1000X. The objective lenses on most laboratory-grade compound microscopes will have magnifications of 10X (low power), 40X (high dry), and 100X (oil immersion).

Objectives

Upon completion of the exercise students should be able to:

1. Identify the parts of the microscope and the function of each component.
2. Distinguish between magnification and resolution.
3. Calculate total magnification.
4. Demonstrate proper care and handling of microscopes.
5. Focus the microscope properly.

Materials

compound light microscope
immersion oil
lens paper
lens cleaner
prepared slides

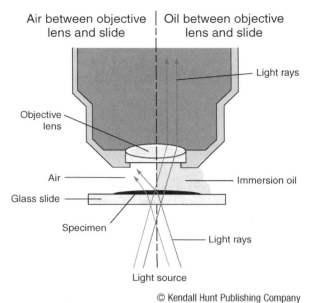

Air between objective lens and slide | Oil between objective lens and slide

- Light rays
- Objective lens
- Air
- Immersion oil
- Glass slide
- Specimen
- Light rays
- Light source

© Kendall Hunt Publishing Company

Figure 1.1. Immersion oil reduces loss of light as it has the same refractive index as glass.

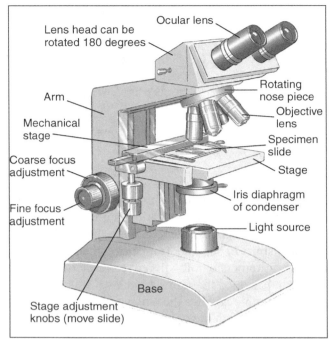

- Lens head can be rotated 180 degrees
- Ocular lens
- Arm
- Rotating nose piece
- Mechanical stage
- Objective lens
- Coarse focus adjustment
- Specimen slide
- Stage
- Fine focus adjustment
- Iris diaphragm of condenser
- Light source
- Base
- Stage adjustment knobs (move slide)

© Kendall Hunt Publishing Company

Figure 1.2. Typical compound microscope.

Methods

✓1. Obtain a microscope from the cabinet making sure to carry it with two hands, one on the arm and one under the base.

✓2. Make sure the stage is lowered all the way using the coarse focus knob. Make sure the rotating nose-piece is turned so that the 10X (low power) objective is in position over the aperture (opening in stage where light passes).

✓3. Plug in the microscope and turn on the light source.

4. Place your slide on the stage. Use the mechanical stage adjustment knobs to position the slide directly over the light source.

5. Adjust the condenser and diaphragm to produce the optimum level of illumination.

6. While looking through the ocular, turn the coarse knob very slowly until the image comes into view.

7. Bring the image into sharper focus by using the fine adjustment knob.

8. Rotate the nose piece so that the 40 (high dry) objective is in place over the specimen.

9. **DO NOT** use the coarse adjustment knob with the 40X or 100X lenses! The microscope is **parfocal**, which means the image should stay in focus when you change lenses. You may need to use the fine adjustment knob to sharpen the image.

10. You may need to adjust the light when you switch between objectives. The iris diaphragm allows you to do this by gently moving the lever just under the front of the stage.

11. Rotate the nosepiece so that there is no lens over the specimen. Add a single drop of immersion oil onto the slide then slide the 100X (oil immersion) lens into the oil.

12. **DO NOT** use the coarse adjustment knob. You may need to very slowly modify the image with the fine adjustment knob.

13. When you are finished remove the slide from your microscope and clean all the oil off using lens paper and lens cleaner.

14. Starting with the lowest power, clean all objective lenses using lens paper and lens cleaner.

15. Rotate the nosepiece so that the lowest power objective lens is in place over the aperture and lower the stage as far as you can.

16. Turn off the light source and wrap the cord around the base. Cover if a dust cover is available. Return the scope to the cabinet making sure to carry it with two hands, one on the arm and one under the base.

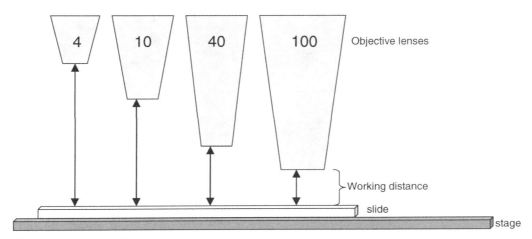

Figure 1.3. Working distance for various objectives.

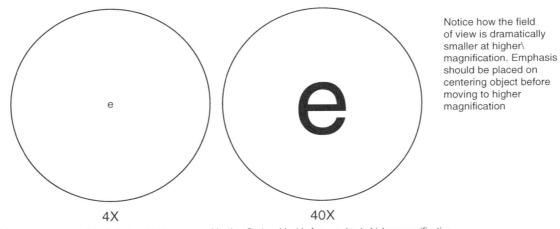

Figure 1.4. Decreasing size of field of view at higher power objective. Center object before moving to higher magnification.

Aseptic Technique

Background

Aseptic techniques are any techniques employed to avoid contamination. In a microbiology laboratory it is often necessary to transfer microbes, most commonly bacteria, from one place to another aseptically. There are multiple precautions utilized to maintain the purity of the cultures that are being manipulated. Initially to prevent contamination, all inoculating instrument (loops and needles) should be sterilized prior to use. This is generally achieved through flaming these objects. All growth media should also be sterilized to ensure an axenic culture (pure culture of organism of interest). Typically growth media is sterilized in an autoclave which utilizes steam under pressure.

Objectives

Upon completion of the exercise students should be able to:

Demonstrate proper methods for the aseptic transfer of bacteria.

Materials

None—Teacher demonstration

Methods

1. All work space should be cleaned with bleach and free of clutter. Light your Bunsen burner. All bacterial transfers and dispensing of media should be done as close to the flame as possible to reduce contamination from airborne microbes.

2. Test tubes are fitted with loose caps so you should always hold the tube and not the cap. Do not lay the tubes on the table or shake them as the contents may spill. Gently mix broth cultures before transfer by rolling them back and forth between your palms.

3. Hold the tube in your nondominant hand (right-handed people should hold the tube in their left hand) and the inoculating loop in your dominant hand.

4. Pass the inoculating loop through the flame of the Bunsen burner. Allow the loop and upper part of the wire to get red hot. This will incinerate any contaminates on the loop. Allow your loop to <u>cool completely</u> before you attempt to pick up cells. A singeing sound indicates you have killed the cells you were attempting to transfer.

5. Grasp the cap of the tube between the ring and pinky fingers and the palm of your dominant hand, being careful not to touch your sterile loop to your hand, the tube, or other surface. Do not lay the cap down and be careful not to touch the opening of the cap to your hand. Pass the opening of the tube through the flame a couple of times. Any spills that occur during a bacterial transfer should immediately be cleaned up with 10% bleach.

6. Insert your sterile loop into the tube and submerge it in the liquid media or touch it to the surface of an agar slant. <u>Be careful not to gouge into the agar</u> if it is a slant culture.

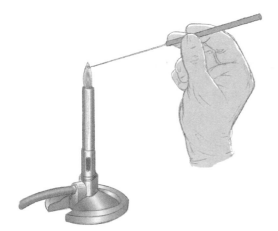

A. Sterilize loop end until red hot. Cool completely before proceeding.

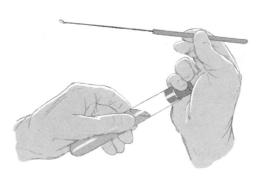

B. Remove stopper from sample tube as shown.

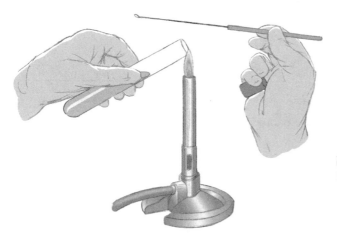

C. Briefly heat tube rim before inserting loop.

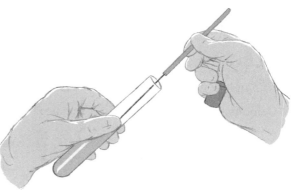

D. Dip loop into sample culture, heat rim again after removing loop, and replace stopper.

E. Tilt up petri dish cover, lightly wipe streaks across surface of agar medium and replace cover.

Figure 2.1. Aseptic transfer of bacteria.

7. Pass the opening of the tube back through the flame, flame the cap opening gently, then replace the cap onto the tube. Put the tube back into the test tube rack. Work quickly as the longer the tube is open to the environment, the more likely contamination will occur.

8. Pick up the tube to be inoculated with your dominant hand. Remove the cap by grasping it between your ring and pinky fingers and your palm. Flame the opening of the tube, then insert your loop. Be careful not to contaminate your loop by allowing it to touch and surface outside the tube. DO NOT heat the loop before you insert it into the growth medium.

9. To inoculate a broth, insert the loop into the fluid and gently mix to remove bacteria from the loop. To inoculate a slant culture, begin at the bottom of the slant and draw the loop across the agar surface in an s-like pattern being careful not to gouge into the agar surface. To inoculate an agar plate, the cover should be only partially removed to form a clam shell-like opening and the inoculating loop should be drawn across the agar surface in an s-like pattern being careful not to gouge into the agar surface. To inoculate a stab culture, an inoculating needle is used to insert the bacterium directly into the agar.

10. Flame the top of the tube and the cap opening. (Agar plates are never flamed but should be only partially opened to allow inoculation.) Replace the cap onto the tube and put it back in the rack. Flame your loop, allowing both the loop and upper part of the wire to get red hot.

11. Most bacterial cultures will be incubated at 37°C for 24 to 48 hours. It is important to label each tube or plate with your name, the date, and what bacteria it contains. Plates should be labeled on the bottom (agar side not removable lid) as they will be stored upside down to reduce condensation on the agar surface.

12. When you have finished your transfer, clean your work space with bleach and wash your hands.

Exercise 3 Ubiquity

Background

Microorganisms are ubiquitous, meaning they are found everywhere. Microbes fill every available ecological niche and some are capable of existing in environmental conditions once believed too harsh to support life. The distribution of microbes within a particular environment is effected by nutrient availability as well as various physical factors, such as temperature, pH and water availability. Within these microbial communities, bacteria have the widest distribution displaying an immense degree of morphological and physiological diversity. Microbes play important roles in agriculture, nutrient recycling, sewage processing, commercial food production, and industry. They are also part of the normal function of the human body. Bacteria and fungi exist as resident flora and outnumber human cells by 10:1. Some microbes are important owing to their ability to cause infectious disease. Bacteria and fungi are too small to be seen individually with the naked eye, but when grown on a solid surface they will clump together and form visible colonies. These colonies can be described in terms of shape, margin (edge of the colony), elevation (side view), texture (wet, shiny, opaque) and pigment (color). Distinctively different colonies indicate different species of microbes. It is assumed that each colony represents the descendents of a single bacterial cell as it reproduces asexually, primarily through binary fission. Fungal colonies tend to be larger with different colors from top or bottom view and may be fuzzy due to spore forming structures.

Objectives

Upon completion of the exercise students should be able to:

1. Demonstrate the ubiquity of microbes in environmental samples.

2. Distinguish between organisms based on colonial appearance.

3. Describe colonial growth characteristics.

4. Recognize growth patterns in broth cultures.

Materials

> 1 plate of Tryptic Soy Agar (TSA)
> 1 tube of Tryptic Soy Broth (TSB)
> 2 sterile cotton swabs

Methods

1. Label the bottom of your plate along the outer edge with group name, section number, and environmental source you decide to sample.

2. Aseptically moisten the sterile swab in the TSB (express excess fluid by pressing swab to inside of tube).

3. Rub the moistened swab over the environmental sample of your choice then streak the swab gently over the TSA surface <u>being careful not to gouge the agar</u>. Be sure that you spread the sample evenly and uniformly so that you cover the entire surface of the plate. Roll the swab along as you spread the sample. You should streak the plate and then turn it clockwise one half turn, then repeat the process. Do this at least twice. Remember to only open the agar plate partially to avoid airborne contaminates.

4. Repeat the sampling process using the second swab, then Place the swab into the TSB tube. You may have to break off the end of the swab to get the top back on.

5. Incubate the plates and broths at 37°C for 24 hours.

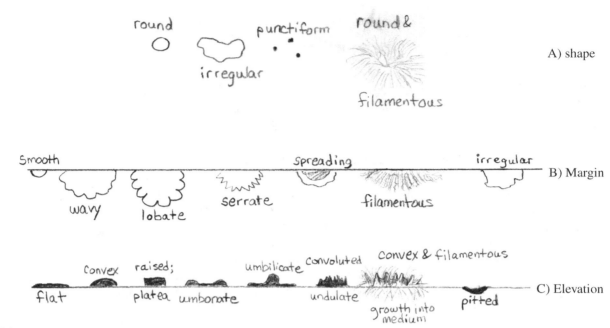

Figure 3.1. Possible colonial shapes (A), margins (B), and elevations (C).

See **Figure 3.2** on page 105.

Pure Culture Techniques

Background

Bacteria typically exist in nature in multispecies communities. It is rare for an independent single species to occur outside of a laboratory setting. It is vital to be able to isolate organisms into a pure culture, one containing a single species, in order to study the morphological and physiological characteristics of an organism. For example, the ability to determine the causative agents of a specific infectious disease hinges on isolating the suspected culprit in a pure culture. Robert Koch, a German physician, was instrumental in developing the first techniques for obtaining pure cultures. Currently multiple methods are available for generating a pure single species culture. One of the most commonly used is the streak plate method. This easy-to-perform method assumes each cell in a mixed culture will give rise to a single pure colony when grown on a semi-solid medium.

Objectives

Upon completion of the exercise students should be able to:

1. Isolate pure cultures from a mixed culture.

2. Utilize the quadrant method to perform a streak plate.

3. Generate bacterial subcultures.

4. Determine the existence of a pure culture using a differential stain.

Materials

    1 TSA plate
    mixed broth culture of *Micrococcus luteus* and *Escherichia coli*
    2 TSA slants

Methods

First Lab Period

1. Label a TSA plate with your group name, section number, and streak plate.

2. Using aseptic technique, obtain a loopful of organisms from the broth culture and lightly streak back and forth through quadrant 1, being careful not to gouge into the agar. Remember to only open the plate enough to successfully inoculate the agar. Close the lid as soon as possible and flame your loop.

3. Rotate the plate 90 degrees, open it slightly, and lightly streak back and forth through quadrant 2 being careful to pull the loop through area 1 several times. Close the plate and flame your loop.

4. Rotate the plate 90 degrees, open it slightly, and lightly streak back and forth through quadrant 3 being careful to pull the loop through area 2 several times. Close the lid and flame your loop.

5. Rotate the plate 90 degrees, open it slightly, and lightly streak back and forth through quadrant 4 being careful to pull the loop through area three several times. Continue the streak out into the center of the plate this time being careful not to come into contact with any of the previously streaked areas. Close the lid and flame your loop.

6. Incubate inverted plates at 37°C for 24 hours.

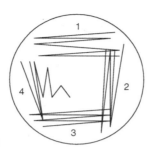

Figure 4.1. Quadrant streak method.

See **Figure 4.2** on page 106.

Second Lab Period

1. Record your observations from your streak plate and answer the related questions on the laboratory report.

2. Label two TSA slants with your group name and section number. Label one M. luetus and one E. coli.

3. Select an isolated yellow colony from your streak plate and aseptically transfer it to the TSA slant labeled M. leutus. Inoculate the slant by starting at the bottom and drawing your loop upward in an s-like pattern across the surface being careful not to gouge into the agar.

4. Repeat the process using an isolated white colony from your streak plate and transfer it to the tube labeled E. coli.

5. Incubate the tubes at 37°C for 24 hours.

Third Lab Period

Perform a gram stain from each subculture to determine if they are in fact single species pure cultures.

5 Smear Preparation, Simple and Negative Stain Techniques

Background

Bacteria can be distinguished from each other in part based on cell morphology (shape) and arrangement. The three common bacterial shapes are coccus (round), bacillus (rod), and spiral (spirillum, spirochete, or vibrio). Depending on the plane of cell division, these cells may remain attached to each other in distinctive patterns. When describing cells one would indicate both the morphology and arrangement. A pair of cells is referred to as having a diplo arrangement. Therefore a pair of round cells would be described as diplococci and a pair of bacilli as diplobacilli. See Figure 5.1 for other possible cell arrangements.

In order to visualize the morphology and arrangements of tiny transparent bacterial cells, a thin smear is prepared and then stained. Smear preparation often involves heat-fixing a bacterial sample to a microscope slide. The addition of heat will kill the cells and adheres them to the surface of the slide so they are not dislodged during the staining process. A danger of using heat to fix a specimen is that cells may become damaged or their cellular structures may become distorted.

Bacteria have multiple negatively charged components so simple stains (single color stains used to illustrate morphology and arrangement) are cationic, positively charged, dyes that are classified as basic dyes. These dyes are attracted to the negative nature of the bacterial cell and bind to the cell so that it takes on the color of the dye and becomes visible using microscopy.

Negative stain techniques employ acidic dyes that are anionic, negatively charged. These dyes are repelled by the negative nature of the bacterial cell and will bind to and stain the slide instead. Organisms will show up as light cells against a dark background. These smears are not heat-fixed or washed so the bacterial cell size and structural components are more easily viewable. Heating may cause the background stain to crack and obstruct the cell features. This technique is also useful in staining cells that do not easily absorb the positive basic dyes.

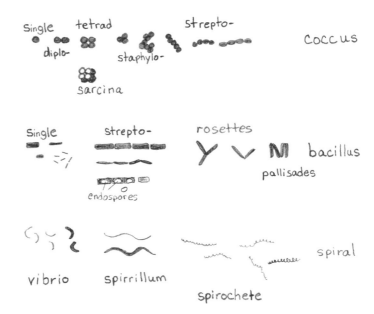

Figure 5.1. Common bacterial cell shapes and arrangements.

Objectives

Upon completion of the exercise students should be able to:

1. Demonstrate good aseptic technique.
2. Utilize the proper procedure for preparing a heat-fixed smear.
3. Demonstrate proper simple stain techniques.
4. Demonstrate negative stain techniques.
5. Distinguish between basic bacterial shapes and arrangements.

Materials

clean microscope slide
slide holder (clothespin)
staining rack
bibulous paper
wash bottle (distilled water)
crystal violet
nigrosin
slant culture of *Staphylococcus aureus*
broth culture of *Bacillus subtilis*

Methods

Smear Preparation

1. All work space should be cleaned with bleach and free of clutter. Light your Bunsen burner. All bacterial transfers should be done as close to the flame as possible in an effort to reduce contamination from airborne microbes.
2. Place a single loopful of water onto the center of your microscope slide. Be careful not to use too much water as this will result in a longer-than-necessary drying time.
3. Using proper aseptic technique, transfer a loopful of the *Staphylococcus aureus* culture into the water on your slide and spread the sample around trying to get a uniform distribution. Be careful not to put too much bacteria on the slide as the resulting smear will be too thick to stain properly. Be careful not to gouge into the agar when removing the bacteria.
4. Allow your smear to completely air dry. Be careful not to blow on or fan the smear so you do not aerosolize the bacteria.
5. Once the smear is completely dry, pick up the slide with the clothespin and pass it through the flame of your Bunsen burner 3 or 4 times. Be careful not to overheat the slide as this will distort the bacterial cells.

   * If you are preparing a smear from a liquid culture, eliminate step 2 and place 2 or 3 loopfuls of the liquid culture directly onto the microscope slide and spread the specimen out.*

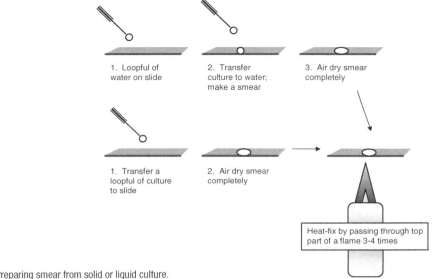

Figure 5.2. Preparing smear from solid or liquid culture.

1. Loopful of water on slide
2. Transfer culture to water; make a smear
3. Air dry smear completely

1. Transfer a loopful of culture to slide
2. Air dry smear completely

Heat-fix by passing through top part of a flame 3-4 times

Simple Stain Technique

1. Place the slide with heat-fixed smear onto the stain rack and cover the surface with crystal violet for 60 seconds.

2. Holding the slide at an angle, gently wash the surface with distilled water for 30 seconds.

3. Insert the slide into a book of bibulous paper and gently blot excess water from the slide. <u>Be careful not to press to hard and break your slide</u>.

4. Examine your slide with a compound microscope. Use the correct technique to move from low power to high power and finally to the oil immersion objective. Record the bacterial cell shape and arrangement you observe.

5. Discard your slide in the metal can on your benchtop. Clean all objectives on your microscope starting with the low power and moving to the high and finally the oil immersion using <u>only lens cleaner and lens paper</u>.

   * **Repeat the steps above using the *Bacillus subtilis* culture.**

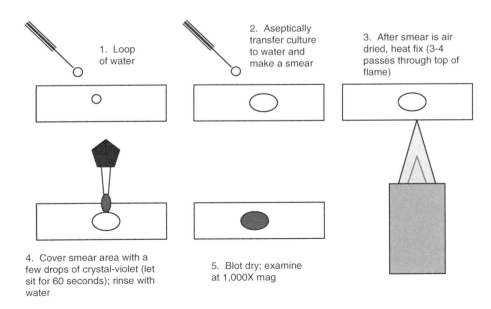

1. Loop of water
2. Aseptically transfer culture to water and make a smear
3. After smear is air dried, heat fix (3-4 passes through top of flame)
4. Cover smear area with a few drops of crystal-violet (let sit for 60 seconds); rinse with water
5. Blot dry; examine at 1,000X mag

Figure 5.3. Simple stain procedure.

Negative Stain Technique

1. Prepare a smear using a loopful of nigrosin instead of a loopful of water. Use *Bacillus subtilis* to prepare the smear. Using a clean microscope slide held at a 45 degree angle, gently push to spread the sample uniformly across the slide. Allow the smear to <u>completely air dry</u>. This smear should <u>not be heat fixed or washed</u>!

2. Examine your slide with a compound microscope. Use the correct technique to move from low power to high power and finally to the oil immersion objective. Record the bacterial cell shape and arrangement you observe.

3. Discard of your slide in the metal can on your benchtop. Clean all objectives on your microscope starting with the low power and moving to the high and finally the oil immersion using <u>only lens cleaner and lens paper</u>.

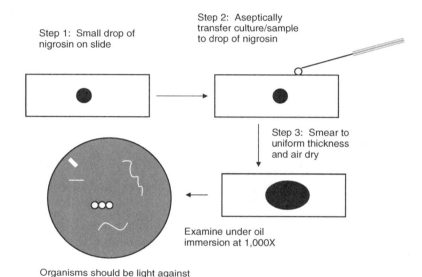

Step 1: Small drop of nigrosin on slide

Step 2: Aseptically transfer culture/sample to drop of nigrosin

Step 3: Smear to uniform thickness and air dry

Examine under oil immersion at 1,000X

Organisms should be light against a dark background

Figure 5.4. Negative stain procedure.

See **Figure 5.5** on page 106.

LABORATORY REPORT

1. Draw what you observed under oil immersion for each organism's simple stain (use colored pencils). Indicate the <u>morphology</u> and <u>arrangement</u> of cells observed for each organism.

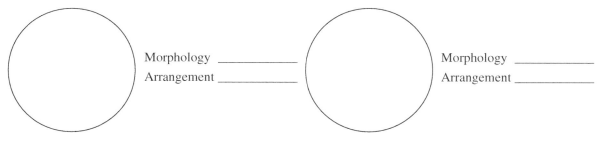

Staphylococcus aureus		*Bacillus subtilis*

Morphology _____ Arrangement _____

Morphology _____ Arrangement _____

2. How do you vary the technique of preparing smears from solid and liquid cultures?

_____

_____

3. What is the purpose of heat fixing a smear?

_____

_____

4. Fill in the appropriate cell morphology and arrangement for the following descriptions:

   Round cells in a chain: _____

   Round cells in a packet of four: _____

   Single comma-shaped cells: _____

   A pair of rod-shaped cells: _____

5. If bacteria are unicellular organisms, why are all organisms you observed not all individual cells?

_____

_____

6. What type of dyes work best for simple stains? What property of these dyes causes them to bind to bacterial cell structures? _____

_____

_____

7. Draw what you observed under oil immersion for your negative stain (use colored pencils). Indicate the <u>morphology</u> and <u>arrangement</u> of all cells observed.

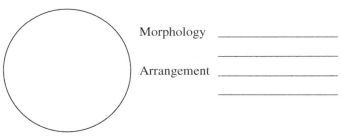

Morphology _____

_____

Arrangement _____

_____

Bacillus subtilis

8. Explain how the stains used for a negative stain differ from those used in a simple stain. How does this affect the result of the stain? _____

_____

_____

9. Why was the heat fixing step not used for the negative stain?

_____

_____

10. Explain the difference in the appearance the organisms when you used a simple and a negative stain technique.

_____

_____

6 Capsule and Spore Structural Stain Techniques

Background

Bacteria are simplistic single-celled organisms; therefore, many species may share a common cell morphology and/or arrangement. Structural stains that detect specific physical or chemical characteristics of a cell may be useful in removing some of the ambiguity of bacterial identification. Many of these stains may also be referred to as *differential stains*. One such structure is a bacterial capsule, a form of glycocalyx; it's typically composed of extracellular layers of sticky polysaccharides or, less commonly, layers of polypeptides or in some instances both. This structure allows bacteria to escape from phagocytic cells, resist desiccation, and aids them in attachment. Capsules can even be used in some cases to distinguish between pathogenic and nonpathogenic strains of a particular bacterial species. Due to their non-ionic nature, capsules can be difficult to stain. Observation of capsules generally involves staining both the cell and background while leaving the capsule as an unstained region around the cell. Congo red, an acidic dye, can be used to stain the background red. However, with the addition of Maneval's stain, which contains acetic acid lowering the pH, the background will shift from red to blue. Maneval's stain also contains a chemical fixative, a basic dye that stains the cells pink and phenol which increases penetration of the stain. This stain technique results in pink cells contrasted by a blue background. The capsules will be visible as colorless regions around the cells.

A less common structure that can be detected through staining is the bacterial endospore. These are dormant structures that provide protection for cells exposed to harsh environmental conditions. Bacterial endospores are not true reproductive spores because a vegetative cell, one actively growing and metabolizing, will produce a single spore then die. These "resting bodies" do provide protection from excessive heat, cold, changes in pH, radiation, and various chemicals. The resistant nature of endospores makes it difficult for them to absorb stains. Malachite green is steamed into the spores and then the vegetative cells are counterstained with safranin which will replace the primary malachite green stain. Many species of bacteria that produce these types of spores are pathogenic and are responsible for diseases such as anthrax, tetanus, and botulism. Spore location and shape vary among species of bacteria. Endospores are typically characterized by shape as terminal, subterminal, or central and common shapes are round, oval, or cylindrical.

Objectives

Upon completion of the exercise students should be able to:

1. Effectively communicate the importance of structural stains in bacterial identification.
2. Demonstrate proper capsule stain techniques.
3. Demonstrate proper endospore stain techniques.
4. Display the technical skills necessary to microscopically differentiate between vegetative cells and endospores.

Materials

    clean microscope slides
    slide holder (clothespin)
    staining rack

electric hot plate
250 ml beaker
bibulous paper
wash bottle (distilled water)
crystal violet
congo red
maneval's stain
malachite green
safranin
slant culture of *Klebsiella pneumoniae*
slant culture of *Bacillus subtilis*

Methods

Capsule Stain Technique

1. Place one drop of 1% Congo red solution on a clean glass slide.

2. Add one loopful of *Klebsiella pneumoniae* and, using you loop, spread it around gently on the slide to form a smear see exercise 5 for smear procedure).

3. Do not heat fix. Allow the suspension to air dry completely.

4. Cover the smear with Maneval's stain for 2 minutes.

5. Hold the slide at an angle to allow the excess stain to drain away. Do not wash the slide with water.

6. Allow the slide to air dry. Do not use the book of bibulous paper to blot the slide dry.

7. Examine your slide with a compound microscope. Use the correct techniques to move from low power to high power and finally to the oil immersion objective. Record your observations.

8. Discard your slide in the metal can on your benchtop. Clean all objectives on your microscope starting with the low power and moving to the high and finally the oil immersion using only lens cleaner and lens paper.

See **Figure 6.1** on page 106.

Endospore Stain Technique

1. Prepare a heat-fixed smear of *Bacillus subtilis* (see exercise 5 for smear procedure).

2. Place the smear on a rack over a beaker of boiling water. Cover the smear with a small piece of bibulous paper and saturate it with Malachite green stain. Steam the smear and stain for 5 minutes. Be careful not to let all the stain evaporate. Add more stain as needed to keep the bibulous paper moist. Be careful not to inhale the fumes as Malachite green can be toxic.

3. Remove smear from heat and allow the slide to cool completely. Remove the bibulous paper and holding the slide at an angle, gently wash the surface with distilled water for 30 seconds.

4. Counterstain by covering the smear with safranin for 30 seconds.

5. Holding the slide at an angle, gently wash the surface with distilled water for 30 seconds.

6. Insert the slide into a book of bibulous paper and gently blot excess water from the slide. Be careful not to press too hard and break your slide.

7. Examine your slide with a compound microscope. Use the correct technique to move from low power to high power and finally to the oil immersion objective. Record the bacterial cell shape and arrangement you observe.

8. Discard your slide in the metal can on your benchtop. Clean all objectives on your microscope starting with the low power and moving to the high and finally the oil immersion using only lens cleaner and lens paper.

See **Figures 6.2** and **6.3** on page 106.

LABORATORY REPORT

1. Draw what you observed under oil immersion for your capsule stain (use colored pencils). Indicate the <u>morphology</u> and <u>arrangement</u> of cells observed.

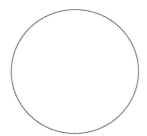

Morphology _____
Arrangement _____

Klebsiella pneumoniae

2. What are three benefits bacteria gain from having a capsule? It allows bacteria to escape from phagocytic cells, resist desiccation, and aids in attachment

3. Why was it not necessary to heat fix the smear for the capsule stain? _____
_____
_____

4. How is it that the capsules remained unstained when both the background and cells were stained?
_____
_____
_____

5. Draw what you observed under oil immersion for your endospore stain (use colored pencils). Indicate the <u>morphology</u> of all <u>cells</u> observed as well as the <u>position</u> and <u>shape</u> of <u>internal spores</u>.

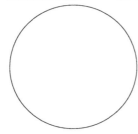

Cell Morphology _____
Spore Position _____
Spore Shape _____

Bacillus subtilis

6. What function did the bibulous paper serve during the steaming required for the primary stain (malachite green)? _____

_____

_____

7. You are given a bacterium that forms a capsule but not an endospore. You are asked to perform a capsule stain but you mistakenly follow the procedure for an endospore stain. How will your result be affected by this mistake?

_____

8. What is the purpose of steaming during the primary stain (malachite green) step of an endospore stain?

_____

9. How would your endospore stain result be affected if you forget to steam the primary stain?

_____

_____

10. Why are endospores not considered to be true reproductive spores? _____

_____

_____

7 Gram and Acid Fast Differential Stain Techniques

Background

The Gram stain, developed by Han Christian Gram in the late 1800s, is one of the most commonly used differential bacterial stains. This stain is often the first step in bacterial identification. Bacteria are grouped into two main categories based on their Gram stain reaction. Gram positive bacterial cells stain purple and gram negative bacterial cells will stain pinkish red. This stain is based on detecting variations in the chemical composition of the bacterial cell wall. Gram positive bacterial cell walls are thick and rigid with many layers of the complex polysaccharide peptidoglycan with interwoven teichoic acids. Gram negative cell walls have little or no peptidoglycan and lack teichoic acids, but they do possess an outer lipopolysaccharide membrane which is not seen in Gram positive cell walls. The primary stain, crystal violet, reacts with the mordant, iodine, to form large crystal-like complexes that are hard to remove from the cell. Teichoic acids of the Gram positive cell wall will react with these complexes and help them be retained in the cell. In addition, the thick layers of peptidoglycan are less susceptible to the organic solvent, ethyl alcohol, than the ethanol soluble lipids of the Gram negative outer membrane. Therefore Gram positive cells will retain the purple color of the crystal violet while the primary stain is almost completely removed from the Gram negative cells. Gram negative cells will instead retain only the safranin counterstain.

Determining the Gram reaction for bacteria isolated from a clinical specimen is vital in prescribing the appropriate antibiotic as many compounds are more effective on one group or the other. Some bacterial cells are considered to be Gram variable. These are Gram positive cells that do not contain consistent amounts of teichoic acids in their cell walls and may give a Gram negative reaction. This may be based on genetics which can code for less teichoic acid than the average Gram positive cell or it could be due to the age of the culture. Older cultures may contain cells that have reduced teichoic acid levels so that some cells may stain Gram negative and some stain Gram positive.

Bacteria of the *Mycobacterium* and *Norcardia* genera have an additional cell wall component called *mycolic acid*. This thick waxy layer of lipids makes these bacteria resistant to water-based stains like those used in a Gram stain. Carbolfuchsin, which contains phenol, is used to penetrate through this waxy substance. The classically used Ziehl-Neelsen method utilizes heat to facilitate the penetration of the stain into the cells. Alternately the Kinyoun method uses a stain with an increased concentration of both carbolfuchsin and phenol to eliminate the need for heat. When *Mycobacterium* and *Norcardia* cells are treated with acid alcohol they will retain the red color of the primary stain. Bacterial cells that lack mycolic acid will release the red stain and can only be visualized when counterstained.

Because *Mycobacterium* and *Norcardia* retain the primary stain even when treated with acid alcohol, they are referred to as acid-fast bacteria. All other bacteria that do not retain the primary stain are considered to be non-acid fast. This is an important stain technique as species of *Mycobacterium* are the causative agents of tuberculosis and leprosy while *Norcardia* species are responsible for a tissue-destroying disease of the hands and feet in addition to pulmonary disease. The thick mycolic acid layer in the cell walls of these bacteria help them to resist environmental pressures like heat, cold, and desiccation as well as many disinfectants and antibiotics.

Objectives

Upon completion of the exercise students should be able to:

1. Effectively communicate the importance of differential stains in bacterial identification.
2. Describe the cellular basis for differential stain reactions.
3. Demonstrate proper Gram stain techniques.
4. Display the technical skills necessary to microscopically differentiate between gram positive and gram negative cells.
5. Demonstrate proper acid fast stain techniques.
6. Display the technical skills necessary to microscopically differentiate between acid fast and non-acid fast cells.

Materials

clean microscope slides
slide holder (clothespin)
staining rack
electric hot plate
250 ml beaker
bibulous paper
wash bottle (distilled water)
70% Ethyl alcohol
acid alcohol
crystal violet
safranin
Gram's iodine
carbolfuchsin
methylene blue
slant culture of *Escherichia coli*
slant culture of *Staphylococcus aureus*
slant culture of *Mycobacterium smegmatis*

Methods

Gram Stain Technique

1. Prepare a heat-fixed smear with *Escherichia coli* (see exercise 5 for smear procedure).
2. Place the slide onto the stain rack and cover the surface with crystal violet for 1 minute.
3. Holding the slide at an angle, gently wash the surface with distilled water for 5 seconds.
4. Cover the surface with Gram's iodine for 30 seconds.
5. Holding the slide at an angle, decolorize with 70% ethyl alcohol for 5–10 seconds depending on how thick your smear is. <u>Be careful not to over-decolorize the smear</u>. This is the critical step for a gram stain and over decolorizing can result in a false gram negative reaction.
6. Stop the decolorizing by holding the slide at an angle and gently washing the surface with distilled water for 10 seconds.
7. Counterstain the smear by covering the surface with safranin for 30 seconds.
8. Holding the slide at an angle, gently wash the surface with distilled water for 5 seconds.
9. Insert the slide into a book of bibulous paper and gently blot excess water from the slide. <u>Be careful not to press too hard and break your slide</u>.

10. Examine your slide with a compound microscope. Use the correct technique to move from low power to high power and finally to the oil immersion objective. Record your observations.

11. Discard your slide in the metal can on your benchtop. Clean all objectives on your microscope starting with the low power and moving to the high and finally the oil immersion using only lens cleaner and lens paper.

*Repeat steps above using the **Staphylococcus aureus** culture.

See **Figure 7.1** on page 107.

Table 7.1 Common mistakes of Gram stains.

	Steps of Gram-stain	Gram-positive	Gram-negative
	1. Crystal-violet	Purple	Purple
	2. Iodine	Purple	Purple
	3. Decolorizer	Purple	Colorless
	4. Safranin	**Purple**	**Red**
	* Forget to add iodine	Red	Red
*Common mistakes	* Overdecolorize	Red	Red
	* Underdecolorize	Purple	Purple

See **Figures 7.2** and **7.3** on page 107.

Acid Fast Stain Technique

1. Prepare a heat-fixed smear with a mix of both *Mycobacterium smegmatis* and *Staphylococcus aureus* (see exercise 5 for smear procedure).

2. Place the slide on a rack over a beaker of boiling water. Cover the smear with a small piece of bibulous paper and saturate it with carbolfuchsin stain. Steam the smear and stain for 5 minutes. Be careful not to let all the stain evaporate. Add more stain as needed to keep the bibulous paper moist. Be careful not to inhale the fumes because the phenol in the carbolfuchsin stain can be toxic, causing irritation to the eyes and mucus membranes.

3. Remove the slide from the heat and allow it to cool completely. Remove the bibulous paper and holding the slide at an angle, gently wash the surface with distilled water for 30 seconds. Be very gentle as the bacteria may not adhere well to the glass slide.

4. Holding the slide at an angle, decolorize with acid alcohol for 5-10 seconds depending on how thick your smear is.

5. Stop the decolorizing by holding the slide at an angle and gently washing the surface with distilled water for 20 seconds. Be very gentle as the bacteria may not adhere well to the glass slide.

6. Counterstain by covering the smear with methylene blue for 30 seconds.

7. Holding the slide at an angle, gently wash the surface with distilled water for 5 seconds. Be very gentle as the bacteria may not adhere well to the glass slide.

8. Insert the slide into a book of bibulous paper and gently blot excess water from the slide. Be careful not to press to hard and break your slide.

9. Examine your slide with a compound microscope. Use the correct technique to move from low power to high power and finally to the oil immersion objective. Record the bacterial cell shape and arrangement you observe.

10. Discard your slide in the metal can on your benchtop. Clean all objectives on your microscope starting with the low power and moving to the high and finally the oil immersion using only lens cleaner and lens paper.

See **Figure 7.4** on page 107.

Exercise 8 **Oxygen Requirements**

Background

Bacteria can be categorized based on their oxygen requirements. The majority of bacteria either utilize oxygen for metabolic processes or they at least tolerate its presence. There are a few bacteria that are actually destroyed by oxygen. The atmosphere contains around 21% oxygen, in addition to other gases like nitrogen, carbon dioxide and hydrogen. Obligate aerobes are bacteria that require oxygen close to this atmospheric level as it is a necessary component of their energy-producing pathways. Microaerophilic bacteria also require oxygen but can only tolerate lower concentrations (typically between 5% to 10%). Anaerobic bacteria can be classified as facultative, aerotolerant, or obligate. Facultative anaerobes will utilize oxygen when it is available but have anaerobic pathways that they employ in its absence. Aerotolerant anaerobes on the other hand do not use oxygen for metabolism but can tolerate its presence without sustaining cellular damage. Truly obligate anaerobes not only do not utilize oxygen but are actually destroyed by it.

The sensitivity of anaerobes to oxygen lies in their inability to produce specific protective enzymes. Multiple toxic by-products are released as oxygen is metabolized. One such product, hydrogen peroxide, is rapidly broken down in aerobic and facultative anaerobic bacteria by the catalase enzyme. This enzyme converts hydrogen peroxide into water and molecular oxygen. Other toxic products of oxygen such as super-oxides and singlet oxygen exist and require additional neutralizing enzymes like superoxide dimutase (SOD). The activity of SOD actually converts superoxide ions into hydrogen peroxide that must then be eliminated. Aerotolerant bacteria have SOD but typically lack catalase. They have alternate enzymes to remove the damaging hydrogen peroxide. Obligate anaerobes have none of these enzyme systems and are therefore destroyed by the toxic forms of oxygen.

Specialized media that reduces the level of atmospheric oxygen can be utilized to determine an organism's oxygen requirements. One such media, thioglycolate broth, contains strong reducing agents like sodium thioglycollate and thioglycollic acid that are rich in sulfhydryl groups. These compounds will reduce the oxygen to water successfully lowering the oxygen levels in the medium allowing for the growth of anaerobes. There is also often a dye, resazurin, added that will serve as a colorimetric indicator of oxygen (colorless when oxygen is absent but pink when it is present). A GasPak system can be used to create a low oxygen, high carbon dioxide environment for incubating plates. This sealed chamber system employs a chemical reaction to release hydrogen gas, which combines with oxygen generating water, and carbon dioxide. A different colorimetric indicator, methylene blue strip, is used with this system to ensure an anaerobic environment exists (strip will fade from blue to colorless as oxygen is removed). Specialized anaerobic agar, which contains a reducing agent, may also be used to further encourage the growth of obligate anaerobes.

Objectives

Upon completion of the exercise students should be able to:

1. Differentiate between bacteria based on oxygen requirements.

2. Effectively utilize reducing media and an anaerobic chamber to cultivate anaerobic bacteria.

3. Detect the production of the protective catalyase enzyme.

Materials

    3 tubes of fluid thioglycollate medium (FTM)
    1 tryptic soy agar (TSA) plate
    1 Brewer's anaerobic agar plate
    1 slant culture of *Escherichia coli*
    1 slant culture of *Bacillus subtilis*
    1 FTM culture of *Clostridium sporogenes*
    1 GasPak chamber
    3 GasPak sachets

Methods

First Laboratory Period

1. Label the TSA plate with your group name and section number.

2. Using aseptic technique, make a single line streak across the top of the plate with the *E. coli,* being careful not to gouge into the agar. Remember to open the plate only enough to successfully inoculate the agar. Close the lid as soon as possible and flame your loop.

3. Using aseptic technique, make a single line streak across the middle of the plate with the *B. subtilis,* being careful not to gouge into the agar.

4. Using aseptic technique, make a single line streak across the bottom of the plate with the *C. sporogenes,* being careful not to gouge into the agar.

5. Be careful to label each streak with the appropriate organism (see diagram below).

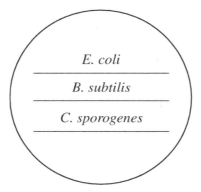

6. Repeat the process with the Brewer's anaerobic agar plate. Then place the inverted plate in a GasPak anaerobic chamber. <u>Do not</u> put the TSA plates into the chamber.

7. Incubate all plates at 37°C for 24 hours.

---

8. Label each FTM tube with your group name and section number. Label one tube *E. coli,* then inoculate it with the *E. coil.* Be careful handling the FTM to avoid allowing any unwanted oxygen into the tube. If more than the top 20% of the media is pink, the media must be boiled for a few minutes to drive off the unwanted oxygen.

9. Label the second tube *B. subtilis,* then inoculate it with the *B. subtilis.*

10. Label the second tube *C. sporogenes,* then inoculate it with the *C. sporogenes.*

11. Incubate the inoculated tubes at 37°C for 24 hours.

Second Laboratory Period

1. Observe and record the growth for both the aerobic and anaerobic plates.

2. Place 2–3 drops of 3% hydrogen peroxide solution onto visible colonies for each organism on the aerobically grown plate. The production of bubbles indicate catalase production.

3. Observe and record the growth for each organism in the FTM tubes.

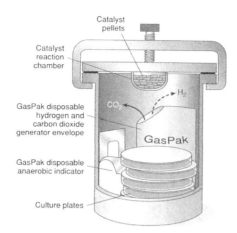

Figure 8.1. GasPak anaerobic growth chamber.

LABORATORY REPORT

1. Record the growth patterns for both the aerobic and anaerobic plates in the table below. Record no growth, some growth, or heavy growth for each organism and positive or negative for catalase production.

Organism	Aerobic Growth	Anaerobic Growth	Catalase
E. coli	little growth	Little growth	Positive
B. subtilis	A lot of growth	A lot of growth	Positive
C. sporogenes	No growth	No growth	Negative

2. What causes the formation of bubbles if catalase is present? The catalase decomposes hydrogen peroxide into water and oxyge and the oxygen produces bubbles

3. Explain how a GasPak chamber works to create an anaerobic environment. _____

4. Record the growth patterns for each organism in the FTM.

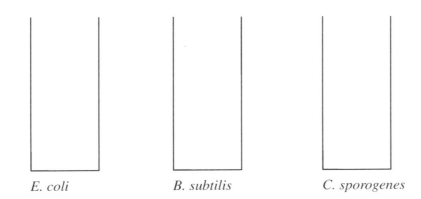

E. coli   B. subtilis   C. sporogenes

5. What is the purpose of adding thioglycollate to the medium? What is the function of the resazurin? oxygen The purpose is to see how much oxygen is needed for bacteria. It helps increase cell viability by using dye

6. Based on the growth patterns observed for each plate and for the FTM cultures, and on catalase production make a determination about each organism's oxygen requirements.

   E. coli   _____

   B. subtilis   _____

   C. sporogenes   _____

7. Explain the difference between an aerotolerant and a microaerophilic organism.

Microphilic organisms need oxygen while aerotolerant does not

8. Explain the difference between an aerotolerant organism and a facultative anaerobe.

Facwative anerobe use oxygen and aerotolerant organism cannot use oxygen.

9 Water Availability

Background

Water is essential for all forms of life as it is a required component of all cellular processes. Osmosis is the passive process of water diffusing from areas of low solute concentration to areas with a higher solute concentration. In an isotonic environment, the solute concentration inside the cell mirrors that of the environment so there is no net movement of water. That is not to say that water is not entering or exiting the cell, but that there is no net gain or loss of water so there is no negative impact on cellular function. When a cell is exposed to a hypotonic environment, one with an excess of solutes inside the cell relative to the external environment, the cell will take on water. Bacterial cells are typically resistant to the osmotic pressure exerted on the cell due to their rigid cell wall. High salt environments have an inhibitory effect on the growth of most bacteria as this creates a hypertonic environment where water is lost to the environment. This may force bacterial cells to undergo plasmolysis where the cell membrane will shrink away from the surrounding cell wall. Loss of water will also cause a decrease in metabolic activity.

Water activity (A_w) is a measure of the water available in a given environment. The water activity of an environment can be dramatically reduced with the addition of salts or sugars. Bacteria can be categorized based on their tolerance to low water activity. Typically bacteria require a water activity of 0.9 or higher (~3% salinity or less) for full metabolic function. Halophiles are organisms that require high salt concentrations (15% to 30%) for growth. They have specific cellular components that require high levels of salinity to function properly. More often than not these organisms fall into the domain Archaea. However, some bacteria are halotolerant, which means that while they do not require high salt conditions they can survive such osmotic pressure. These halotolerant organisms are common inhabitants of the skin and mucus membranes.

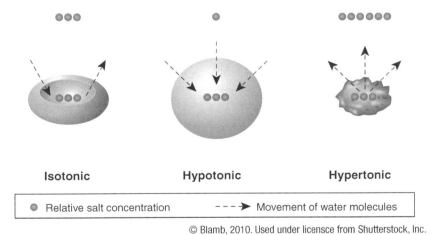

Isotonic   Hypotonic   Hypertonic

● Relative salt concentration   - - - ➤ Movement of water molecules

© Blamb, 2010. Used under licensce from Shutterstock, Inc.

Figure 9.1. Effect of osmotic pressure on a red blood cell. Bacteria are somewhat protected from these effects due to their rigid cell wall.

Objectives

Upon completion of the exercise students should be able to:

1. Effectively communicate the concept of water activity.
2. Evaluate the effects of osmotic pressure and water availability on bacterial growth.
3. Make determinations about halotolerance of specific organisms.

Materials

1 tryptic soy agar (TSA) plate with 0.5% NaCl
1 tryptic soy agar (TSA) plate with 5% NaCl
1 tryptic soy agar (TSA) plate with 10% NaCl
1 slant culture of *Escherichia coli*
1 slant culture of *Bacillus subtilis*
1 FTM culture of *Staphylococcus aureus*

Methods

1. Label each TSA plate with your group name, section number, and salt concentration.
2. Using aseptic technique, make a single line streak across the top of each plate with E. coli, being careful not to gouge into the agar. Remember to only open the plate enough to successfully inoculate the agar. Close the lid as soon as possible and flame your loop.
3. Using aseptic technique, make a single line streak across the middle of each plate with B. subtilis, being careful not to gouge into the agar.
4. Using aseptic technique, make a single line streak across the bottom of each plate with Staphylococcus aureus, being careful not to gouge into the agar.
5. Be careful to label each streak with the appropriate organism (see diagram below).

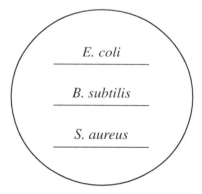

6. Incubate all plates at 37°C for 24 hours.
7. Observe and record the growth for all plates.

LABORATORY REPORT

1. Record the growth patterns for each organism at each NaCl concentration in the table below. Record no growth, some growth, or heavy growth for each.

Organism	0.5% NaCl	5% NaCl	10% NaCl
E. coli	Medium growth	Medium growth	High growth
B. subtilis	Little growth	Little growth	High growth
S. aureus	Little growth	Medium growth	High growth

2. Based on the data above, what conclusion can you make about each organism's tolerance to increasing salt concentrations (no salt tolerance, halotolerant, halophile)?

   E. Coli ___halotolerant___

   B. Subtilis ___halotole no salt tolerance___

   S. aureus ___halophile___

3. Which of these organisms do you think could successfully colonize the skin and the mucus membranes?
   I think the E.coli and B.subtilis could successfully colonize the skin and the mucus membrane

4. Generally speaking, what makes bacteria more resistant to hypotonic environment than other microbes like protozoa? Bacteria have cell walls that resist osmotic pressure that causes cells to swell.

5. What is the effect of a hypertonic environment on most bacterial cells? Water is lost from the cell which can lead to eventual death (plesmolysis)

6. What is meant by water availability and why is it important to normal cellular function?
   Water availability is pretty self explanatory, its the availability of water. Its important because water helps the movement of waste and the stability within a cell

Exercise 10 Soil Microbiology— Population Counts

Background

Biodiversity is a measure of the variety of organisms, including the number of species and genetic variation within a species, in an ecosystem. Soil may appear to be relatively inert or lifeless, but most soils are in fact full of life. The major soil organisms include microbes like bacteria and fungi as well as invertebrates such as arthropods. A single teaspoon of soil can contain thousands of species of microbes and maybe even billions of individuals. Some microbes may be plant or animal pathogens, but many are beneficial to an ecosystem. Microbes are necessary for decomposition of organic debris and play vital roles in nutrient cycles like the nitrogen and sulfur cycles. Soil microbes are also a source of compounds such as antibiotics on which humans have come to depend.

Factors that affect soil formation also affect the biodiversity. These include climate, texture, and vegetation. The chemical (pH, nutrient availability, and salinity) and physical conditions (water availability and temperature) of the soil also influence the inhabitants. Collectively these factors influence the number and types of organisms that can inhabit the soil. It is important to realize that soil microbes may actually influence the very factors that impact their ability to populate a habitat.

There are several common techniques for measuring the number of soil microbes. One such method is the plate count which involves dilutions of a soil sample being plated using various growth media to select for certain organisms. Serial dilutions are used to reduce the number of microbes present to a countable number. Plates are incubated allowing for the growth of microbes into visual colonies which are then counted. It is assumed each colony originates from a single bacterial cell, but this is often not the case. Colonies are referred to as a CFU or colony forming unit that allows for the more likely possibility of a colony originating from a small clump of cells. Plate counts tend to underestimate microbes. Studies show that only 1% to 10% of soil organism have been isolated and grown in culture. This is typically due to the fact that some microbes may have very specific growth requirements.

Objectives

Upon completion of the exercise students should be able to:

1. Evaluate the number and types of microbes in soil and explain their roles.

2. Perform serial dilutions and viable plate counts.

3. Describe what is meant by colony forming unit and how to correctly count plates.

4. Calculate the total number of organisms per gram of sample.

Materials

3 sterile 99ml water blanks
4 Petri plates
1.1 ml pipettes
soil sample
Only one of the following per group:
    1 bottle (50ml) of molten Tryptic Soy Agar (TSA)—bacterial growth medium

1 bottle (50ml) of molten Potato Dextrose Agar (PDA)—fungal growth medium
1 bottle (50ml) of molten Glycerol Yeast Extract Agar (GYEA)—Actinomycete growth medium

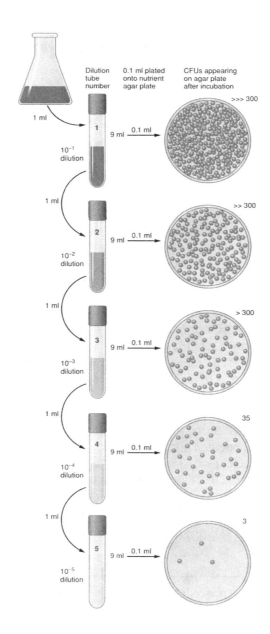

Figure 10.1. Serial dilutions for a standard plate count.

Methods

1. Label the 4 Petri plates according to the dilutions required for your organism (based on the media you received):

Bacteria	Fungi	Actinomycetes
1:10,000	1:1000	1:100
1:100,000	1:10,000	1:1000
1:1,000,000	1:100,000	1:10,000
1:10,000,000	1:1,000,000	1:100,000

2. Label the 99ml water blanks A, B, and C.

3. Add 1 gram of soil to blank A (1:100), shake hard for 3 minutes or until the soil is dissolved. Transfer 1ml from blank A into blank B (1:10,000) and shake hard for 1 minute. Transfer 1ml from blank B into blank C (1:1,000,000) and shake hard for 1 minute. If you have been assigned the Actinomycetes, you will not need blank C (1:1,000,000).

4. Distribute the proper amounts of water from the blanks to the plates to achieve the final dilutions needed for your organism. Use the following guidelines:

1:100	1.0 ml blank A
1:1,000	0.1 ml blank A
1:10,000	1.0 ml blank B
1:100,000	0.1 ml blank B
1:1,000,000	1.0 ml blank C
1:10,000,000	0.1 ml blank C

5. Retrieve a bottle of media and allow it to cool just until you can hold it comfortably in your hand. Pour just enough media to cover the bottom (approximately ¼ of the bottle) into each plate then swirl the mixture gently. Do not swirl the media too vigorously. It should not be allowed to splash up onto the lid of the dish. Allow the media to solidify. <u>Do not swirl the media once it begins to solidify.</u>

6. Place the plates into the incubator for 24 hours. Count only plates with an acceptable number of CFUs (30–300) and record your data, then answer the corresponding questions.

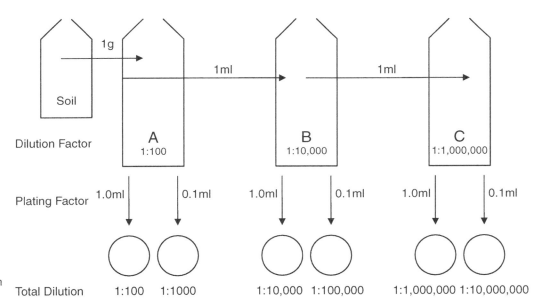

Figure 10.2. Serial dilution and plating of soil sample.

LABORATORY REPORT

1. To determine the number of organisms in your original soil sample you must account for your final dilution [(number of colonies (dilution factor ÷ plating factor)]. Select a countable plate (30–300 colonies) and calculate the number of CFUs per gram of soil. Record your results below.

Organism	CFUs per Plate	Final Dilution (Dilution ÷ Plating factor)	CFUs per Gram of Soil
Bacteria			
Actinomycetes			
Fungi			

Questions:

2. Based on the CFUs/gram, what conclusions can you draw from your data?
   _____

3. What is meant by CFU? _____
   _____
   _____

4. Why might the number of bacteria in your soil sample be higher than the number you estimated from your plate count? _____
   _____
   _____

5. What other types of organisms (micro or macro) might be present in your soil sample?
   _____
   _____

6. You may see evidence of antibiosis. Describe the appearance of any colonies that seem to be producing an antibiotic compound. What is the response of other colonies on the plate to the presence of this compound? _____
   _____
   _____
   _____

7. What factors may effect the biodiversity of a soil sample?
   _____
   _____

8. Why was it necessary for you to perform serial dilutions of you soil sample?
   _____
   _____

9. Briefly explain the role microbes play in the soil.
   _____
   _____

10. Why was a different media type used for each of the microbes you were attempting to culture?
    _____
    _____

11 Quantification of Microbes in Food

Background

Food microbiology, a branch of microbiology, includes three major areas of concern. These areas are food spoilage, food-borne illness, and food production. Fermented foods are some of the oldest and most widely consumed. Microbial fermentations may provide characteristic flavors, aromas, and textures of certain foods like yogurt and bread. However, microbes may also be responsible for devastating economic losses due to food spoilage. In addition, food-borne illness such as Salmonellosis, Campylobacteriosis, botulism, and Staphylococcal food intoxication affect thousands of people each year. Spoilage bacteria and pathogens can be inadvertently introduced during processing or preparation of foods. This necessitates stringent quality control procedures. The standard plate count is a common method for enumerating bacteria in food samples. This technique involves the use of serial dilutions to reduce the number of bacteria to a countable number. The dilutions are plated using Plate Count Agar which is a media that most common spoilage and pathogenic bacteria present in food can utilize. Plates are incubated to allow for the growth of bacteria into visible colonies which are then counted. Differential media may be used to detect the presence of specific pathogens.

Objectives

Upon completion of the exercise students should be able to:

1. Communicate the relevance of detecting bacteria in foods.
2. Quantify bacteria present in common foods.
3. Perform serial dilutions and viable plate counts.

Materials

    1 sterile 180 ml water blank
    1 sterile 99 ml water blank
    3 Petri plates
    1.1 ml pipettes
    food sample
    1 bottle (50ml) of molten Plate Count Agar
    blender

Methods

1. Label each Petri plate with your group name, food type, and the dilution (1:100, 1:1000, or 1:10,000) that will be plated.
2. Add 20 grams of your food sample to the 180 ml water blank and blend the mixture for 5 minutes. This will provide a 1:10 dilution of your food sample.
3. Distribute 0.1 ml from the blended mixture into the plate labeled 1:100.
4. Transfer 1.0 ml from the blender into the 99 ml water blank to create a 1:1000 dilution of your food sample.

5. Distribute 1.0 ml from 99 ml bottle into the plate labeled 1:1000.

6. Distribute 0.1 ml from 99 ml bottle into the plate labeled 1:10,000.

7. Retrieve a bottle of media and allow it to cool just until you can hold it comfortably in your hand. Pour just enough media to cover the bottom into each plate, then swirl the mixture gently. Do not swirl the media too vigorously. It should not be allowed to splash up onto the lid of the dish. Allow the media to solidify. <u>Do not swirl the media once it has begun to solidify</u>.

8. Place the plates into the incubator for 24 hours. Count only plates with an acceptable number of CFUs (30–300) and record your data then answer the corresponding questions.

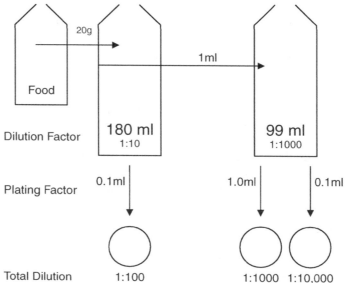

Figure 11.1. Serial dilution and plating of food sample.

LABORATORY REPORT

1. To determine the number of organisms in your original food sample you must account for your final dilution [number of colonies (dilution factor ÷ plating factor)]. Select a countable plate (30–300 colonies) and calculate the number of CFUs per gram of food. Record your results below.

Food Type	CFUs per Plate	Final Dilution (Dilution ÷ Plating factor)	CFUs per 20 grams of Food

Questions:

2. Based on the number of CFUs/gram, what conclusions can you draw from your data? _____
_____

3. Which food type had the highest bacterial count? _____
_____

4. Which food type had the lowest count? _____
_____

5. What factors could account for these results? _____
_____
_____
_____

6. Would this type of plate count be used to assess the safety of fermented foods like yogurt? Why or why not. _____
_____
_____
_____

7. Why was only one type of media used for each food sample? _____
_____

8. In what situation would it be better to use a differential media? _____
_____

9. What would be a practical application for using a standard plate count to assess the number of bacteria in a food sample? _____

_____

10. Was it necessary to do serial dilutions before you plated the sample? Explain why or why not.

_____

_____

Exercise 12 — Evaluation of Disinfectants and Antiseptics

Background

A wide variety of chemicals are utilized in the control of unwanted microbial growth. These anti-microbial chemicals are typically classified as either antiseptics or disinfectants. While they have the common goal of removing or reducing the number of unwanted microbes, their application varies. Disinfectants are anti-microbial chemicals meant for use on inanimate surfaces while antiseptics are considered safe for use on living tissues. It is possible for some compounds to be used in either capacity. Some chemicals are truly sterilizing agents and will destroy all forms of microbial life. Many anti-microbial chemicals are not capable of eliminating all microbes but are useful sanitizing agents (reduce microbial numbers to an acceptable level).

A disk diffusion test is a relatively simple method of assessing the effectiveness of a specific chemical at inhibiting microbial growth. The test involves exposing bacterial inoculated agar to a filter disk impregnated with the chemical, incubating the plate, then looking for an effect on the bacterial growth pattern. If the organism is susceptible to the effect of the chemical, a zone of inhibition (area with no bacterial growth) will be seen around the filter disk. The size of the zone of inhibition indicates the level of effectiveness for a given compound against that organism. Through simple diffusion, the chemical will move away from the filter disk and throughout the agar. A concentration gradient will be established with the highest concentration of the chemical occurring immediately adjacent to the filter disk. Therefore, the farther away from the disk that inhibition of growth occurs, the lower the concentration of chemical responsible for it. The efficacy of a compound may vary depending on the characteristics of the microbe it is being tested against. For this experiment we will test various anti-microbial compounds against the gram positive bacterium, *Staphylococcus aureus*, and *Pseudomonas aeruginosa*, which is gram negative.

Objectives

Upon completion of the exercise students should be able to:

1. Distinguish between an antiseptic and a disinfectant.

2. Perform and interpret a disk diffusion test in order to evaluate the effectiveness of anti-microbial chemicals.

3. Provide reasonable explanations for the any observed variation in microbial susceptibility to anti-microbial chemicals.

Materials

Various antiseptics and disinfectants
2 plates of Tryptic Soy Agar (TSA)
broth culture of *Staphylococcus aureus*
broth culture of *Pseudomonas aeruginosa*
sterile swabs
sterile filter disks
forceps

Methods

1. Label two TSA plates with your group name, section number and the disinfectant/antiseptic you have been assigned. Label one plate *S. aureus* and the other *P. aeruginosa*.

2. Using good aseptic technique, saturate a sterile swab with *S. aureus* then press the swab onto the inner side of the tube to express some of the fluid. Bacteria in a broth culture tend to settle to the bottom. Gently mix the culture before you saturate the swab to re-suspend the bacteria.

3. Inoculate the entire surface of the appropriate TSA plate <u>evenly and uniformly</u> by streaking the swab across the middle of the plate then rotating the plate clockwise one-half turn and streaking again. Rotate the plate clockwise one-half turn a second time and streak it again. Finally streak the swab around the entire outer edge of the plate. <u>Be very careful not to gouge into the agar</u>.

4. Allow at least 3 minutes for the broth to be absorbed into the agar.

5. Flame sterilize your forceps by passing them through a flame quickly two or three times and then select a single sterile filter disk. Expose the disk to your assigned disinfectant/antiseptic by allowing the edge of the disk to touch the surface of the chemical. <u>Be careful not to oversaturate the disk</u>.

6. Place the disk in the center of the TSA plate inoculated with *S. aureus*. <u>Gently tap the disk</u> with the forceps to ensure it does not fall off when you invert the plate. <u>Be careful not to gouge into the agar</u>. Do not relocate the disk once you have placed it on the agar surface.

7. Repeat the process using the same chemical and a TSA plate inoculated with *P. aeruginosa*.

8. Incubate the inverted plates at 37° for 24 hours.

9. Following incubation, measure and record the size of each zone of inhibition (from the center of the disk to the area where growth begins) in millimeters.

See **Figures 12.1** and **12.2** on page 108.

#20 →

#26 → S. aureus

Exercise 13 Fungi

Background

Fungi represent a group of eukaryotic microbes that includes unicellular yeasts and multi-cellular molds. Many people are only familiar with fleshy fungi which produce plant-like reproductive structures called mushrooms. However, fungi are not plants but are instead chemoheterotrophs that may be saprophytic, feed on decaying organic material, or exist as parasites of a living host. Fungal cells do have distinct cell walls, but they are composed of polysaccharides like chitin and not the cellulose seen in plant cell walls. Fungi also lack the tissue differentiation seen in higher plants.

Molds are filamentous organisms consisting of multiple interwoven tubular structures called hyphae. The total mass of an organism's hyphae is referred to as a mycelium. The fungal mycelium is divided into two parts. The vegetative hyphae serve to anchor the fungus and to absorb nutrients from the environment. Aerial hyphae are reproductive in nature and produce asexual spores.

Asexual fungal spores are classified as either sporangiospores, formed within a sac called a sporangium, or conidia, formed in chains externally on a conidiophore. Conidia can take multiple forms including those produced in chains off of a ball-like structure or off fingerlike projections called phialides. Many fungi produce irregular thick-walled spores called Chlamydospores that serve to ensure survival in harsh environmental conditions. Unicellular yeast can be oval or spherical and typically reproduce asexually through budding or fission. Yeasts will behave like bacteria when grown on solid surfaces forming visible multicellar colonies.

Fungi are typically classified into phyla based on their sexual reproductive cycle. There are three types of sexual spores formed by fungi. Fungi in the phylum Zygomycota (conjugation fungi) produce sexual zygospores through the union of nuclear material from two genetically variant hyphae. The Ascomycota (sac fungi) produce haploid sexual spores, ascospores, within a specialized enclosure called an ascus. Sexual basidiospores are produced by members of the Basidomycota (club), which includes fungi that produce common mushrooms and puffballs. These basidiospores are produced externally on a pedestal-like structure called a basidium. Basidia are typically located along the underside of the gills of common mushrooms like *Coprinus*.

Molds play a beneficial role in an ecosystem by recycling nutrients through their decomposition of organic material. Molds are also beneficial to humans in that they are a source of many antibiotic compounds. Certain molds are used in the production of food preservatives like citric acid while many fleshy fungi, mushrooms, may themselves be eaten. Yeasts are facultative anaerobes that play a role in food production through their fermentative abilities. Non-pathogenic species like *Saccharomyces* are utilized in the production of alcoholic beverages like beer and wine as well as bread.

Some fungi are in fact pathogenic to humans or plant resulting in infections called mycoses. These fungi may be dimorphic yeast, capable of two forms of growth, a phenomenon that is typically dependent on environmental conditions. Species of *Candida* will form pseudohyphae under specific environmental conditions which cause budding cells to remain attached to the parent cell. The formation of pseudohyphae allows the yeast to penetrate deeper into body tissues, which results in serious systemic effects on the host.

Objectives

Upon completion of the exercise students should be able to:

1. Describe the basic features of fungi.
2. Differentiate between yeasts and molds.
3. Recognize the various types of sexual and asexual spores produced by fungi.
4. Explain the beneficial and harmful aspects of fungi.

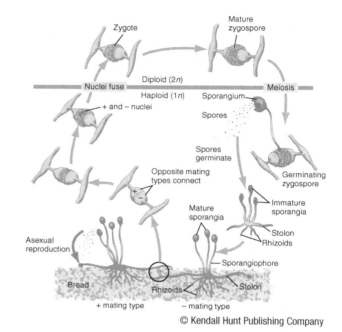

Figure 13.1. Life cycle of a zygomycete showing production of asexual sporangiospores and sexual zygospores.

© Kendall Hunt Publishing Company

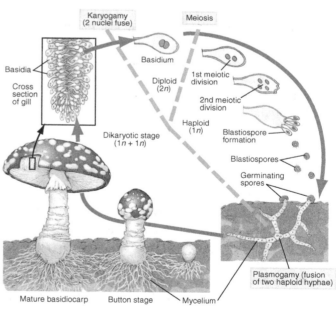

Figure 13.2. Sexual life cycle of basidomycete showing formation of mushroom (fruiting body).

© Kendall Hunt Publishing Company

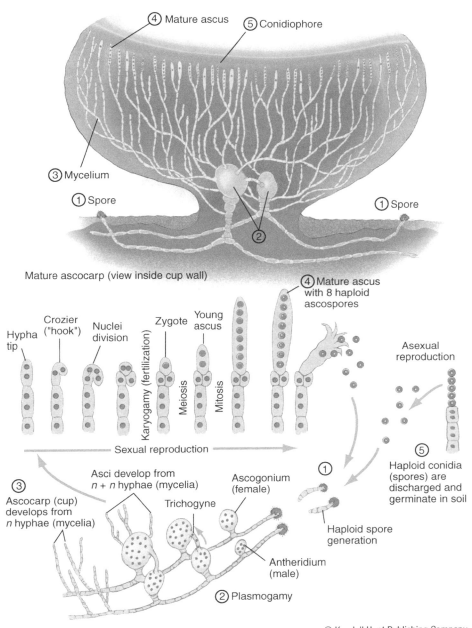

Figure 13.3. Life cycle of ascomycete showing release of asexual spores from conidia and sexual spores from asci.

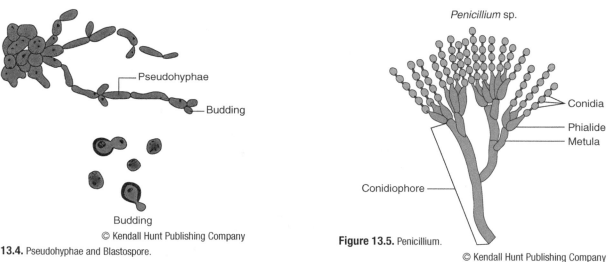

Pseudohyphae

Budding

Budding

© Kendall Hunt Publishing Company

Figure 13.4. Pseudohyphae and Blastospore.

Penicillium sp.

Conidia

Phialide

Metula

Conidiophore

Figure 13.5. Penicillium.

© Kendall Hunt Publishing Company

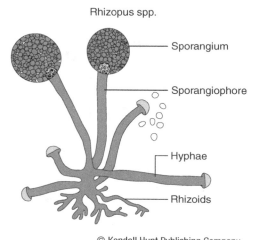

Rhizopus spp.

Sporangium

Sporangiophore

Hyphae

Rhizoids

Figure 13.6. Rhizopus.

© Kendall Hunt Publishing Company

See color images of **Figures 13.4, 13.5,** and **13.6** on page 108.

Materials

Prepared slides of the following:
    Candida yeast
    Penicillium mold
    Mucor mold
    Coprinus mushroom
    Compound light microscope

Methods

1. Obtain a microscope from the cabinet making sure to carry it with two hands, one on the arm and one under the base.

2. Make sure the stage is lowered all the way using the coarse focus knob. Make sure the rotating nose-piece is turned so that the 10X (low power) objective is in position over the aperture (opening in stage where light passes).

3. Plug in the microscope and turn on the light source.

4. Place your slide on the stage. Use the mechanical stage adjustment knobs to position the slide directly over the light source.

5. Adjust the condenser and diaphragm to produce the optimum level of illumination.

6. While looking through the ocular, turn the coarse knob very slowly until the image comes into view.

7. Bring the image into sharper focus by using the fine adjustment knob.

8. Rotate the nosepiece so that the 40 (high dry) objective is in place over the specimen.

9. **DO NOT** use the coarse adjustment knob with the 40X or 100X lenses! The microscope is **parfocal**, which means the image should stay in focus when you change lenses. You may need to use the fine adjustment knob to sharpen the image. You may need to adjust the light when you change between objectives.

10. Record observations for the *Mucor, Penicillium,* and *Coprinus* slides using the 40X high dry objective. For the *Candida* slide, follow steps 11 to 13 to utilize oil immersion.

11. Rotate the nosepiece so that there is no lens over the specimen. Add a single drop of immersion oil onto the slide then slide the 100X (oil immersion) lens into the oil.

12. **DO NOT** use the coarse adjustment knob. You may need to very slowly modify the image with the fine adjustment knob.

13. When you are finished, remove the slide from your microscope and clean all the oil off using lens paper and lens cleaner.

14. Starting with the lowest power, clean all objective lenses using lens paper and lens cleaner.

15. Rotate the nosepiece so that the lowest power objective lens is in place over the aperture and lower the stage as far as you can.

16. Turn off the light source and wrap the cord around the base. Cover if a dust cover is available. Return the scope to the cabinet making sure to carry it with two hands, one on the arm and one under the base.

LABORATORY REPORT

1. Draw observations for the prepared fungal slides below. Utilize your colored pencils. Record observations about any visible spore types or fungal morphology.

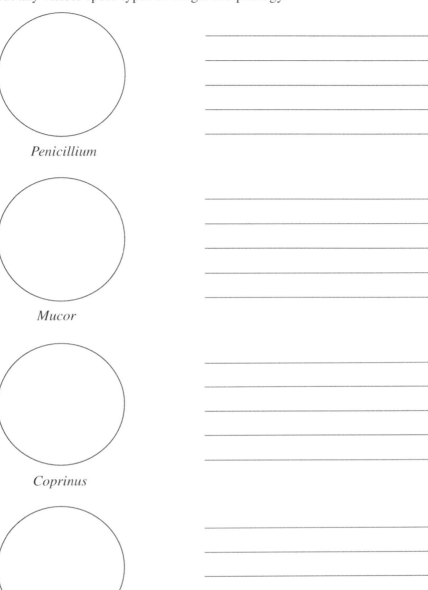

Penicillium

_____

_____

_____

_____

_____

Mucor

_____

_____

_____

_____

_____

Coprinus

_____

_____

_____

_____

_____

Candida

_____

_____

_____

_____

_____

2. The tubular structures forming the filaments of molds are called _____ while the entire visible mass of fungal growth is referred to as a _____.

3. Differentiate between asexual sporangiospores and conidia. _____
_____

4. Differentiate between yeast, molds and fleshy fungi. _____
_____
_____

5. What is meant by dimorphic fungi? What controls this behavior and what is the potential benefit to the fungus? _____
_____
_____

6. List three beneficial aspects of fungi. _____
_____
_____

7. Yeasts typically reproduce by _____ but some species may form _____ if buds do not separate completely.

8. List the different types of sexual spores produces by fungi. _____
_____

9. Fungi from the phylum _____ produce classic mushrooms and puffballs to distribute sexual spores while members of the phylum _____ produce sexual spores within an enclosure called an ascus.

10. List three reasons why fungi are not classified as plants. _____
_____
_____

Exercise 14 Bacteriophages and Plaque Formation

Background

Bacteriophages (phages) are viruses that infect bacteria. The viral DNA is injected through the cell wall directly into the host cell, where it directs the production and release of hundreds of progeny phages. Virulent (lytic) phages destroy their bacterial host when they have completed their replication cycle. The release of the newly replicated phages ruptures the host cell (killing it) and they are then free to infect more bacterial host cells. When bacteria are grown on a semisolid surface, the infection and subsequent lysis of host cells will produce clear areas called plaques. The production of these plaques indicates the susceptibility of a bacterium to a particular phage. Phages, like all viruses, have a specific host range. The ability of a phage to bind to and invade a host cell is dependant on the phage recognizing and binding to specific host receptor sites, such as lipopolysaccharides or teichoic acids, on the cell's surface. Bacteria that lack the appropriate receptor site are immune to infection from that particular phage.

International literature contains several hundred reports dating from the 1920s on phage therapy. This involves the application of phages for treating infectious bacterial diseases. Phages are very specific, affects the targeted bacterium only; therefore, damage to the host is highly unlikely. It is also unlikely for the phage to have an impact on the host's normal flora. Phage therapy would therefore reduce the risk of developing secondary infections commonly associated with broad spectrum antibiotic therapy.

Objectives

Upon completion of the exercise students should be able to:

1. Describe the specificity of the phage-bacterium relationship.

2. Describe the lytic action of a virulent phage.

3. Identify a phage plaque.

4. Explain the potential benefits of phage therapy.

Materials

suspension of T_4 bacteriophage
broth culture of *Escherichia coli B*
broth culture of *Proteus vulgaris*
4 plates of Tryptic Soy Agar (TSA)

Methods

1. Label the bottom 1 TSA plate *E. coli* B; label the bottom 1 TSA plate *E. coli* B with phage; label the bottom 1 TSA plate *P. vulgaris;* and the final TSA plate *P. vulgaris* with phage.

2. Inoculate each plate with the appropriate bacteria. To ensure a complete lawn of bacterial growth the plates should be inoculated with a multiple streak method using a sterile swab. Dip the swab into the broth culture of bacteria and then swab the surface of the agar carefully in a streak pattern. Make a 1/4 turn

of the plate and repeat the streak. Repeat the process then make a circular streak around the outer rim of the agar against the edge of the plate. To avoid contamination, open the plate only slightly (like a clam shell) while performing the streak. **Be sure to use a new sterile swab for the each inoculation**. Discard the used swabs into disinfectant, not the trash can.

3. Inoculate the plates labeled "with phage" with two loopfuls of T4 phage. Using your inoculating loop make a single streak down the center of the plates.

4. Incubate plates at 37°C for 48 hours, then store a room temperature until the next lab meeting.

5. Observe the plates for a clear line against a lawn of bacterial growth. The presence of a line indicates the formation of plaques where the phage infected and destroyed the susceptible bacterial host cells. Record your results on the laboratory report.

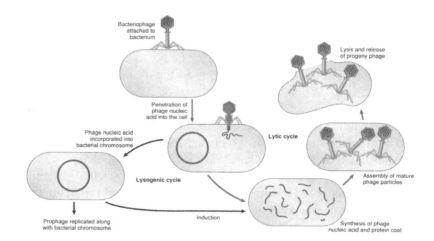

Figure 14.1. Life cycle of a lytic bacteriophage.

LABORATORY REPORT

Draw the growth patterns for each of your plates.

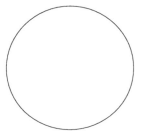

E. coli and phage

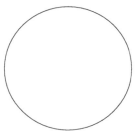

E.coli only

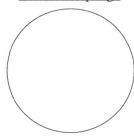

P. vulgaris and phage

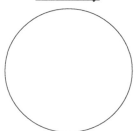

P. vulgaris only

1. Which bacterial species appears to be susceptible to the T4 phage? _____

2. What would influence a bacteria's susceptibility to a particular phage? _____
   _____

3. Explain how this variation in susceptibility increases the beneficial application of phage therapy for the treatment of bacterial disease. _____
   _____

4. What is meant by the term *lytic* and how does it relate to plaque formation? _____
   _____
   _____

Exercise 15 Oxidation, Fermentation, and Hydrolysis Reactions

Background

Bacteria can be difficult to differentiate due to limited distinctive morphological features. Bacteriologists use a wide variety of physiological (biochemical) characteristics to aid in the identification of bacteria. These types of biochemical tests are based on the activity of enzymes used to catalyze a wide range of metabolic reactions. Many bacteria release enzymes into the environment (exoenzymes) where they catalyze hydrolysis reactions. These reactions involve the break down of large polymers like starch, proteins or lipids into sub-units that can be incorporated into the bacterial cell. Other tests are directed at detecting an organism's ability to perform aerobic or anaerobic respiration. Such tests may involve detecting oxidase enzymes needed for aerobic respiration, oxygen neutralizing enzymes like catalase, or certain by-products of anaerobic respiration. An alternate anaerobic pathway is fermentation. Multiple tests are available to determine which sugars a bacterium is capable of fermenting as well as tests directed at detecting products of specific fermentative pathways. Many of the available tests utilize differential media or indicator reagents which will detect the presence of particular metabolic products. Specific tests can be preformed as a series to differentiate between bacteria with similar morphological and physiological features. IMViC reactions are employed to distinguish between members of the *Enterobacteriacae* family (enterics). It is especially useful in differentiating between groups of enterics called coliforms, which are used as indicators of fecal contamination. The name IMViC indicates the four tests involved: indole production, Methyl-red test, Voges-Proskauer test, and citrate utilization.

Objectives

Upon completion of the exercise students should be able to:

1. Effectively explain the use of metabolic characteristics to identify bacteria.
2. Recognize the activity of extra-cellular hydrolytic enzymes.
3. Recognize the fermentation of simple sugars, milk sugars, and protein by bacteria.
4. Recognize the degradation of specific amino acids and urea by bacteria.
5. Explain the procedure for and benefit of an IMViC test.
6. Describe reactions used to detect aerobic or anaerobic respiration in bacteria.
7. Detect the presence of protective enzyme in bacteria.

Materials

Media:
2 glucose fermentation broths with Durham tube
2 lactose fermentation broths with Durham tube
4 MRVP broths
2 Simmons Citrate slants
2 Nitrate broths with Durham tube
2 TSA plates

1 Starch agar plate
2 Tryptone broths
2 Phenylalanine agar slants
2 Urea broths
4 Sims agar deeps

Cultures (slant of each):
Escherichia coli
Enterobacter aerogenes
Proteus vulgaris

Bacillus subtilis
Staphylococcus aureus
Pseudomonas aeruginosa

Reagents:
Methyl red
3% Hydrogen peroxide
Ferric chloride
Alpha-naphthol (VPA)
Potassium hydroxide (VPB)
Sulfanilic acid (nitrate A)

N,N dimethyl-alpha naphthylamine (Nitrate B)
Zinc powder
Gram's iodine
N,N-dimethyl-p-phenylenediamine
hydrochloride (Oxidase reagent)
p-Dimethylaminobenzaldehyde hydrochloride
(Kovac's reagent)

Methods

First Laboratory Period (inoculate media for biochemical tests)

Fermentation Tests (determines if organism is capable of various type of fermentation)
1. Label one glucose broth, one lactose broth, one citrate slant, and two MRVP broths with your group information, section number, and the name of the control organism listed in the table in the laboratory results section. One MRVP broth should be labeled MR and the other VP. (For example, MR tube will be labeled group name, section number, MR, and *E. coli* and the VP tube group name, section number, VP, and *E. aerogenes*.)

2. Label one glucose broth, one lactose broth, one citrate slant and two MRVP broths with your group information, section number and the name of an unknown organism of your choice (any organism not listed as the control organism in the table in the laboratory results section). One MRVP broth should be labeled MR and the other VP. (For example, MR tube will be labeled group name, section number, MR, and any organism except *E. coli* and the VP tube group name, section number, VP, and any organism except *E. aerogenes*.)

3. Aseptically inoculate each tube with the organism indicated on the label.

Oxidative Tests (determines if organism carries out respiration, aerobic or anaerobic)
1. Label one nitrate broth with your group information, section number and the name of the control organism listed in the table in the laboratory results section. (For example, one nitrate tube will be labeled group name, section, and *E. coli*.)

2. Label one nitrate broth with your group information, section number and the name of an unknown organism of your choice (any organism not listed as the control organism in the table in the laboratory results section). (For example, one nitrate tube will be labeled group name, section, and any organism other than *E. coli*.)

3. Aseptically inoculate each tube with the organism indicated on the label.

4. Label 2 TSA plates with your group name and section number. Label one plate oxidase and one plate catalase. Draw a line down the center of each plate.

5. On one side of the catalase plate write the name of the control organism listed in the table in the laboratory results section. On the opposite side write the name of an unknown organism of your choice (any organism not listed as the control organism in the table in the laboratory results section). Do not use *Proteus* as your unknown. It is a swarming bacterium that will overgrow on your plate.

6. Aseptically inoculate each side of the plate with the organism indicated on the label.

7. Repeat steps 5 and 6 for the oxidase plate. Remember not to use *Proteus* as your unknown.

Hydrolytic Tests (determines bacterial production of hydrolytic/degradative enzymes)

1. Label a starch plate with your group name and section number. Draw a line down the center of each plate.

2. On one side of the starch plate write and the name of the control organism listed in the table in the laboratory results section. On the opposite side write the name of an unknown organism of your choice (any organism not listed as the control organism in the table in the laboratory results section). Do not use *Proteus* as your unknown. It is a swarming bacterium that will overgrow on your plate.

3. Aseptically inoculate each side of the plate with the organism indicated on the label.

4. Label one tryptone broth, one urea broth, and one phenylalanine slant with your group information, section number, and the name of the control organism listed in the table in the laboratory results section.

5. Label one tryptone broth, one urea broth, and one phenylalanine slant with your group information, section number, and the name of an unknown organism of your choice (any organism not listed as the control organism in the table in the laboratory results section).

6. Aseptically inoculate each tube with the organism indicated on the label.

Multi-tube Tests (utilizes media designed to give multiple test results)

1. Label three Sims deeps with your group information, section number, and the name of **one** control organism listed in the table in the laboratory results section.

2. Label one Sims deep with your group information, section number, and the name of an unknown organism of your choice (any organism not listed as a control organism in the table in the laboratory results section).

Second Laboratory Period (examine all inoculated tubes and record results)

Fermentation Tests (determines if organism is capable of various types of fermentation)

1. Examine each glucose and lactose fermentation broth. This media measures the production of gas and acid which are typical by-products of fermentation. If the media has turned from red to yellow this indicates a positive reaction for acid production. If there is a visible bubble in the inverted Durham tube, the test is positive for gas production.

See **Figure 15.1** on page 108.

2. The MRVP media measures the production of specific by-products of fermentative pathways. This will involve two tests and will require the addition of reagents.

3. The methyl red test is used to identify bacteria that use mixed acid fermentation pathways. Add 5 drops of methyl red to the culture and agitate gently. If the liquid turns red immediately that indicates a positive reaction. If no color change occurs then the test is negative.

4. The Vogues-Proskauer test identifies bacteria that produce neutral end products such as 2,3 butanediol. Add 1 ampoule of VPA (alpha-naphthol) reagent and 1 ampoule of VPB (potassium hydroxide) reagent and agitate the tube vigorously. This will result in the 2,3 butanediol being oxidized to acetoin. If the media turns red within 30 minutes the test is positive for acetoin. If after 30 minutes there has been no color change, the test is negative.

See **Figure 15.2a** and **Figure 15.2b** on page 108.

5. Examine each citrate slant. If the organism was able to utilize the citrate as its sole source of carbon then the media will be bright cerulean blue (positive). If the organism lack the appropriate enzyme for the utilization of citrate (citratase) then the media will remain green.

See **Figure 15.3** on page 108.

Oxidative Tests (determines if organism carries out respiration, aerobic, or anaerobic)

1. Examine both the nitrate broths for the presence of a bubble in the Durham tube. A bubble indicates the reduction of nitrate/nitrite to a gaseous end product like N_2O or N_2. This process is called denitrification.

2. Some bacteria partially reduce nitrate to nitrite utilizing among others the enzyme nitrate reductase. Detection of nitrite involves the addition of reagents.

3. If the nitrate broth has no visible bubble, add 3 drops of nitrogen reagent A (sulfanilic acid) and 3 drops of nitrogen reagent B (N,N, dimethyl-alpha naphthylamine). If the media turns dark red immediately, the test is positive for nitrite. If no color change occurs, the test is negative for nitrite production.

4. To confirm that the nitrate in the medium has not been reduced, add a pinch of zinc powder. The zinc will reduce the available nitrate and the media will immediately turn a bright red. If no color develops, your organism may have produced some other product of nitrate reduction like ammonia.

See **Figures 15.4** and **15.5** on page 109.

5. Observe the growth on the TSA plate labeled oxidase. The enzyme oxidase is utilized in the electron transport chain of aerobic respiration. Add 5 drops of oxidase reagent (N,N-dimethyl-p-phenylenediamine hydrochloride) onto colonies on both sides of the TSA plate. Leave the plate open to expose the culture to oxygen while you are waiting for the color to develop. If the colonies turn purple, the organism is positive for the production of oxidase. If no color develops within one minute, the test is negative. If the color develops after one minute, this may be a false positive and should not be recorded as a positive result.

See **Figure 15.6** on page 109.

6. Observe the growth on the TSA plate labeled catalase. The enzyme catalase neutralizes a toxic form of oxygen called hydrogen peroxide. As catalase degrades the hydrogen peroxide water and gaseous oxygen are released. Add 5 drops of 3% hydrogen peroxide onto colonies on both sides of the TSA plate. Leave the plate open to expose the culture to oxygen while you are waiting for the bubbles to develop. If bubbles are produced, the organism is positive for the production of catalase. If no bubbles appear, the test is negative.

See **Figure 15.7** on page 109.

Hydrolytic Tests (determines bacterial production of hydrolytic/degradative enzymes)
1. Observe growth on your starch plate. Bacteria that produce the extra-cellular enzyme amylase are able to degrade starch and utilize the liberated glucose. Cover the colonies on both sides of the plate with iodine. The iodine will bind to the starch and produce a dark color (purple to black). A clear zone will appear around the colonies of bacteria that produce amylase and therefore have degraded the starch indicating a positive reaction.

See **Figure 15.8** on page 109.

2. Examine the urea broths. Bacteria that produce the urease enzyme convert urea into ammonia. The presence of ammonia turns the phenol red indicator bright pink. So if the media is pink the organism is positive for urease production and if there is no color change the test is negative.

See **Figure 15.9** on page 109.

3. Examine the phenylalanine slants. Certain gram negative bacteria are capable of deaminating (removing the amine group) the amino acid phenylalanine producing phenylpyruvic acid and ammonia. This reaction requires the enzyme phenylalanine deaminase. Add several drops of ferric chloride to each slant. If the colonies turn green, the organism is positive for phenylalanine deaminase. If no color develops within 5 minutes, the test is negative.

See **Figure 15.10** on page 110.

4. Examine the tryptone broths. This test indicates the hydrolysis of the amino acid tryptophan into indole, pyruvate, and ammonia. The reaction is catalyzed by the enzyme tryptophanase. Add one indole ampoule which contains Kovac's reagent (p-Dimethylaminobenzaldehyde hydrochloride) to each broth. Do not agitate the tubes. If a red ring forms at the top of the broth the test is positive. If a beige or yellow ring forms, the test is negative.

See **Figure 15.11** on page 110.

Multi-tube Tests (utilizes media designed to give multiple test results)

1. Examine your SIM deeps. As we observed in the previous reaction, indole is a waste product of tryptophan hydrolysis. The SIM deep is a semi-solid medium that not only detects the presence of indole but also tests for sulfur production and motility. This media contains tryptophan but also sulfur containing cysteine and iron salts. As with the previous reaction, the addition of one indole ampule (Kovac's) reagent will produce a red ring on top of the media if indole is present due to the hydrolysis of tryptophan. The hydrolysis of cysteine, catalyzed by the enzyme cysteine dusulfurase, releases hydrogen sulfide which will combine with iron salts forming a visible black precipitate. Motility can be assessed by looking for growth away from the central stab line. You may need to hold the tube up to the light to observe any evidence of motility.

See **Figure 15.12** on page 110.

LABORATORY REPORT

1. Record results for all reactions in the table below. Be sure to indicate the unknown organism used for each test.

Test	Control	Unknown	# per group	Media/reagent	Unknown Reaction (+ or −)
Glucose Fermentation Tube - **no reagent required**	*E. coli*		1 control, 1 unknown	Fermentation tubes	Acid _____ Gas _____
Lactose Fermentation Tube - **no reagent required**	*E. coli*		1 control, 1 unknown	Fermentation tubes	Acid _____ Gas _____
MRVP (MR – mixed acid)	*E. coli*		1 control, 1 unknown	MRVP broth & methyl red	
MRVP (VP – butanediol)	*E. aerogenes*		1 control, 1 unknown	MRVP broth & VP ampoules A & B	
Citrate utilization - **no reagent required**	*E. aerogenes*		1 control, 1 unknown	Simmons Citrate slant	
Oxidase Test	*P. aeruginosa*		1 control 1 unknown	TSA plate and oxidase ampoule	
Catalase production	*S. aureus*		1 control, 1 unknown	TSA plate and 3% hydrogen peroxide	
Nitrate reduction	*E. coli*		1 control, 1 unknown	Nitrate broth and nitrate A and B ampoules	Nitrogen gas _____ Nitrate _____ Other nitrogen products_____
Starch hydrolysis	*B. subtilis*		1 control, 1 unknown	Starch agar plates and Gram's iodine	
Indole Test	*E. coli*		1 control 1 unknown	Tryptone broth; Indole ampoule (Kovac's reagent)	
Urea hydrolysis - **no reagent required**	*P. vulgaris*		1 control, 1 unknown	Urea broth	
Phenylalanine deamination	*P. vulgaris*		1 control, 1 unknown	Phenylalanine slant and ferric chloride	
SIM Test	*S. aureus, E. coli, P. vulgaris*		3 control 1 unknown	SIM deep; Indole ampoule (Kovac's reagent)	Motility ____ H$_2$S ____ Indole ____

2. What does IMViC stand for and how is it useful? _____

_____

3. What is an exoenzyme? Give two examples. _____

_____

4. Indole is produced when the amino acid _____ is broken down.

5. A and B reagents are added to nitrogen broth to detect _____
and zinc powder is added to detect _____.

6. **Match the following reagent with the biochemical test in which it was used. Answers will be used only once.**

_____ iodine	a.	mixed acid fermentation
_____ Kovac's reagent	b.	nitrate reduction
_____ alpha-naphthol	c.	catalase production
_____ ferric chloride	d.	2,3 butanediol fermentation
_____ hydrogen peroxide	e.	phenylalanine deamination
_____ N,N, dimethyl-alpha naphthylamine	f.	starch hydrolysis
_____ methyl red	g.	tryptophan degradation

7. **Match the following media with the biochemical test in which it was used. Answers may be used more than once. Some questions may require more than one answer.**

_____ catalase production	a.	MRVP broth
_____ 2,3 butanediol fermentation	b.	SIM deep
_____ tryptophan degradation	c.	TSA
_____ hydrogen sulfide production	d.	tryptone broth
_____ mixed acid fermentation	e.	phenol red lactose broth
_____ carbohydrate fermentation		

8. **Match the following enzyme with the reaction it catalyzes. Answers will be used only once.**

_____ casein hydrolysis	a.	catalase
_____ breakdown of hydrogen peroxide	b.	urease
_____ starch hydrolysis	c.	cysteine desulfurase
_____ hydrogen sulfide production	d.	oxidase
_____ indole production	e.	caseinase
_____ hydrolysis of urea	f.	tryptophanase
_____ involved in aerobic electron transport	g.	amylase

9. Many biochemical pathways produce gaseous products. List two reactions from this exercise that may have produced gaseous end products and indicate how the presence of the gas was detected. _____
_____
_____

10. Why is SIM media useful in determining motility? How does the inoculation of this media differ from other media used in this exercise? _____
_____
_____

Exercise 16 — Unknown Identification

Unknown Identification

Background

Taxonomy is the science of naming and grouping organisms based on genetic relationships. Classification of multicellular organisms often utilizes morphology and fossil records to form phylogenetic groups. This can be somewhat difficult for bacteria owing to the presence of few morphological forms and no complete fossil record. Species definition varies dramatically between higher organisms and bacteria. For higher organisms the definition of species is based on sexual reproduction and the production of fertile offspring. Bacteria are asexual organisms that form genetic recombinants through methods like conjugation, so the biological species concept is not applicable. A bacterial species is currently defined as a group of prokaryotic cells with a set of unique characteristics.

Microbiologists are often charged with the identification of unknown bacteria. Identification of bacteria is aided by the use of Bergey's *Manual of Systematic Bacteriology*, which exists in several volumes listing all described and isolated bacteria. Various phenotypic and genotypic characteristics have been provided for each identified organism. Organisms are divided into groups based on sharing a common set of these traits. Identification typically begins by examining the Gram stain reaction and morphological characteristics of an organism and progresses to an examination of biochemical characteristics. A flowchart is useful in dividing large numbers of bacteria into smaller groups based on the results of specific tests.

Objectives

Upon completion of the exercise students should be able to:

1. Utilize information and techniques gained from previous laboratory exercises to culture and identify and unknown bacterium.
2. Effectively employ a flowchart in the identification of an organism.

Materials

> 1 unknown bacterial culture
> various stains, media, reagents, and supplies

Methods

1. Begin by inoculating a TSA slant to create a subculture of your unknown for further testing.
2. Perform a Gram stain of your unknown bacterial culture.
3. Based on the Gram reaction and morphology of your bacteria follow the appropriate flowchart on the following pages to determine the identity of your unknown bacterium.

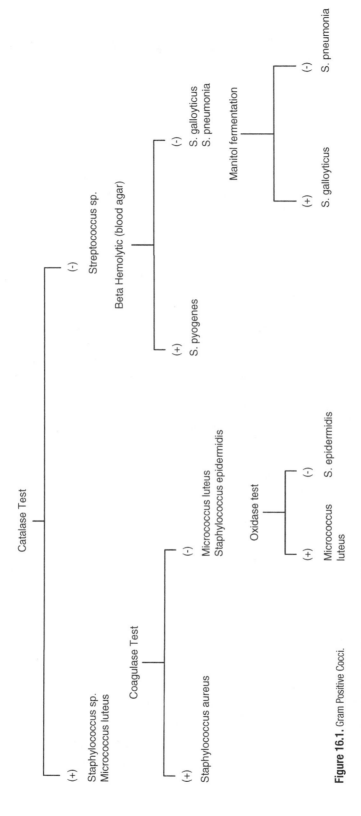

Figure 16.1. Gram Positive Cocci.

See **Figures 16.4** and **16.5** on page 110.

82

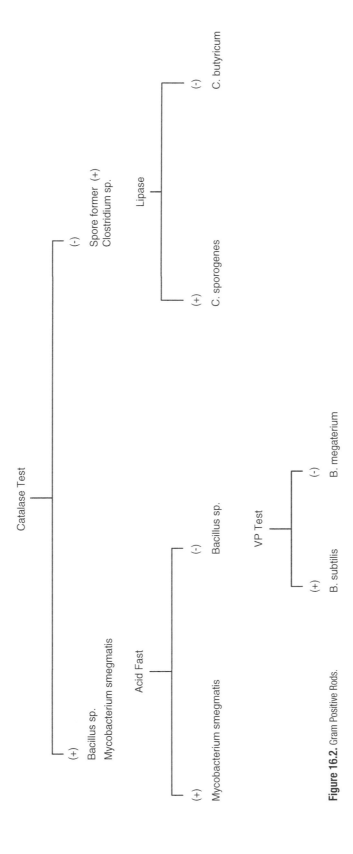

Figure 16.2. Gram Positive Rods.

83

84

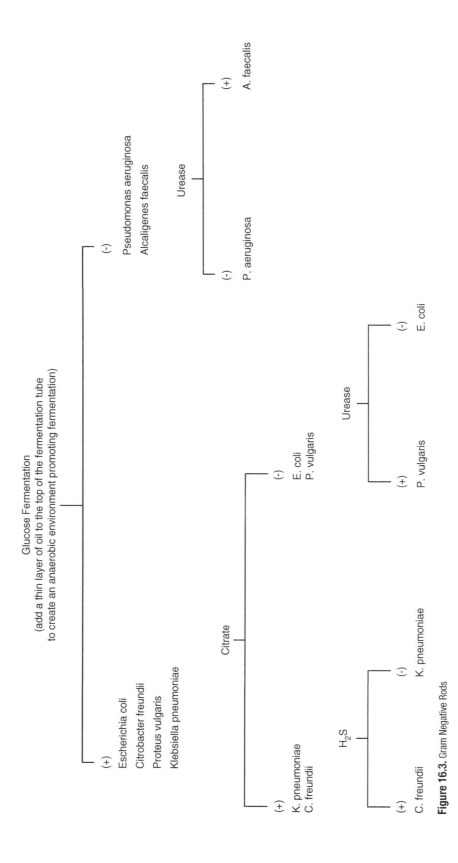

Glucose Fermentation
(add a thin layer of oil to the top of the fermentation tube
to create an anaerobic environment promoting fermentation)

(+)

Escherichia coli
Citrobacter freundii
Proteus vulgaris
Klebsiella pneumoniae

(-)

Pseudomonas aeruginosa
Alcaligenes faecalis

Citrate

(+)

K. pneumoniae
C. freundii

(-)

E. coli
P. vulgaris

Urease

(-)

P. aeruginosa

(+)

A. faecalis

H₂S

(+)

C. freundii

(-)

K. pneumoniae

Urease

(+)

P. vulgaris

(-)

E. coli

Figure 16.3. Gram Negative Rods

17 Unknown Identification Using Box PCR Amplified DNA Fingerprints

Background

Prior to cell division, a bacterial cell makes a copy of its chromosome through a process called deoxyribonucleic acid (DNA) replication. Each DNA molecule is made up of two complementary strands and DNA replication begins with the separation of these two strands. This process requires the action of various enzymes and DNA binding proteins. Each strand of the separated DNA molecule can now serves as template for the production of the missing complementary strand. During DNA replication an enzyme called DNA polymerase adds complementary nucleotides using the single parental strand as a template. This allows for the production of two identical replicas of the original DNA molecule. The process is referred to as semi-conservative replication as each new DNA molecule contain both an original parent strand and a newly synthesized strand.

Polymerase chain reaction (PCR) employs the principles of semi-conservative DNA replication to amplify DNA fragments *in vitro* into billions of copies in a short period of time. This technique was developed in 1983 by Kary Mullis. For PCR amplification, a reaction mixture (about 25-50 µl) is prepared by adding source DNA (the template to be copied), Taq-polymerase (a thermostable DNA polymerase isolated from a thermophilic bacteria called *Thermus aquaticus*), primers (specifically designed short DNA fragments containing sequences complementary to each target region), deoxyrybonucteotides (dNTPs; substrate for Taq-polymerase), PCR buffer (a salt-solution that helps to stabilize the DNA and other reaction components) and MgCl2 (required for activity of Taq-polymerase). The PCR reaction is then run in repeating cycles of heating and cooling in a machine called a thermal cycler. A typical reaction cycle consists of three steps (denaturation, annealing and extension or synthesis). In the first step (denaturation), the two strands of the DNA are separated by heating the reaction mixture to an elevated temperature (typically to around 94°C). In the second step (annealing), the temperature is lowered (typically to around 50-60°C) which allows the primers to bind to specific sites on of each of the separated DNA strands. In third step (extension or synthesis), the temperature is raised (typically to about 72°C) and the Taq-polymerase elongates the primers by adding complementary nucleotides (dNTPs). These three steps are repeated 30-35 times to generate billions of copies of target DNA. The total number of copies can be estimated by the following equation: **Total DNA copies after PCR: Starting number of copies $\times$ 2$^{\text{number of cycle}}$**

Several types of PCR have been developed thus far. In this exercise you will perform a BOX-PCR to amplify fragment(s) of DNA from an unknown bacterium. The principle of BOX PCR relies on the presence of several short interspersed repetitive DNA sequences that occur in prokaryotic genomes. Although their exact function is currently unknown, the presence of these repetitive DNA sequences can be utilized for DNA fingerprinting based identification of bacterial species. Repetitive element sequence-based PCR (rep-PCR) amplifies the region between two repetitive sequences of a bacterial genome. Uniquely designed primers bind interspersed repetitive sequences that are separated by various distances in the bacterial chromosome.

PCR amplification of the sequences between interspersed repeats results in different sized amplification products, which are unique among bacterial species. One such approach, BOX-PCR, uses a single BOX A1R primer (5' CTA CGG CAA GGC GAC GCT GAC G 3') to amplify regions between box sequences in the bacterial genome.

The amplified fragments can be separated through gel electrophoresis. This process involves loading the DNA fragments into wells in a porous agarose gel then using an electrical current to separate the fragments into visible bands. DNA moves through the pores in the agarose gel in response to the current because each fragment has a negative charge due to the phosphate group on the nucleotides. Agarose gel electrophoresis is performed using submerged conditions (gel submerged in a buffer solution). To increase the density of the DNA fragments before loading, the fragments are mixed with a loading buffer (contains high molecular weight compound and bromophenol blue dye) so that they will sink in the well and do not mix with the buffer. The bromophenol blue in the loading buffer helps track the movement of DNA fragments through the gel.

During electrophoresis, smaller DNA molecules move faster than larger molecules. This results in the DNA fragments being separated on the basis of their molecular weight. By running a standard DNA ladder (contains fragments with known molecular weights), the molecular weight of an unknown DNA fragment can be determined. The DNA fragments separated in the gel can be visualized by using an ethidium bromide dye that binds to DNA and fluoresces under UV light. The band patterns (size and number of bands) created through electrophoresis of the amplified repetitive regions are unique among species, thus a species can be identified by examining its band pattern (DNA fingerprint).

Objectives

Upon completion of the exercise students should be able to:

1. Effectively communicate the principles of Polymerase Chain Reaction (PCR) and gel electrophoresis and how they can be applied in the identification of bacterial species.

2. Utilize PCR and electrophoresis techniques to identify an unknown bacterial species.

Materials

Micropipettes and tips
Microfuge tubes
PCR tubes
Thermal cycler
Ice bucket
Vortex
Unknown slant culture
BOX A1R primer
BOX PCR supermix (Contains PCR buffer, Taq polymerase, $MgCl_2$, dNTPs)
H_2O (nuclease free)
Conical flask (50-100 ml)
Agarose
TAE buffer (1X)
Gel casting tray and combs
Electrophoresis chamber and power supply
6X Sample Loading Buffer
DNA ladder
Ethidium bromide (EtBr)

Methods

First Laboratory Period (DNA extraction and PCR amplification)

1. **DNA extraction:** BOX PCR does not require high purity DNA. Bacterial genomic DNA samples for BOX PCR can be prepared by following steps:

   a. Take a bacterial culture grown overnight on a TSA slant.

   b. Take a loopful of bacteria and suspend it in 100 µl of deionized water in a 1.5 ml microfuge tube.

   c. Mix well. Vortex to disperse the cells.

   d. You will need 1.0 µl of this cell suspension (contains template DNA) when you prepare the PCR reaction mix in the next step.

2. **Preparation of PCR reaction mix:** Prepare PCR reaction mix as described in the table below. Keep materials on ice while preparing the reaction mix.

Components	Volume
Nuclease free H_2O	21.0 µl
BOX PCR supermix	25.0 µl
BOX primer	2.0 µl
DNA template (from step 1)	2.0 µl
Total	**50.0 µl**

3. **Run PCR amplification:** Perform PCR amplification in a thermal cycler using following cycling conditions.

   Single step at 95°C for 2 min (Initial denaturation)
   35 cycles of:
       94°C for 3 sec
       92°C for 30 sec
       50°C for 60 sec
       65°C for 8 min
   Single step at 65°C for 8 min (final extension)

4. After completion of the PCR amplification, store your samples at 4°C. You will use this sample to run agarose gel electrophosesis (next lab), to visualize any DNA fragments that were amplified in your sample.

Second Laboratory Period (agarose gel electrophoresis)

Preparing the agarose gel

a. Measure 0.5 g of agarose powder into a 50-100 ml flask. Add 25 ml 1X TAE buffer to the flask. This will make a 1% agarose gel. Different % of agarose gel can be used depending on the size of the DNA sample that you want to separate.

b. Dissolve the agarose by boiling in a microwave until the solution becomes clear (2 or 3 30 sec cycles with mixing between each cycle).

c. Let the solution cool to about 50-55°C, add 1ul of ethidium bromide and mix well (avoid generation of bubbles during mixing).

d. Pour the gel in casting tray with a comb and let it cool until it solidifies (appears white).

e. When gel solidifies, carefully pull out the combs and place the gel in the electrophoresis chamber. Make sure you place the wells toward the negative pole (black), so the DNA will '**Run to Red'** (towards positive pole).

f. Add just enough TAE buffer to cover 2-3 mm over the gel.

Loading the gel

In a fresh PCR tube, mix 1 µl of 6X sample loading buffer and 5 µl of your PCR sample. Using a micro-pipette, carefully transfer 5 µl of each sample into a well in the gel. Similarly, load 5 µl of the DNA ladder standard into one of the wells.

Running the gel

a. Place the lid on electrophoresis chamber and connect the electrode wires to the power supply. Make sure that the positive (red) and negative (black) wires are connected properly.

b. Run the power supply at around 100 volts (**5 volts/cm**). You can see bubbles generating from electrodes and movement of the bromophenol blue tracking dye.

c. When the dye approaches the end of the gel, turn off the power and remove the gel tray.

Visualization of DNA

DNA bands can be visualized using a UV light. DNA fragments that are bound to EtBr in the gel fluoresce as red bands under the UV light. A gel image, like the one illustrates below in figure 17.1, can be acquired using a suitable gel documentation system.

Interpretation of result

Analyze the band pattern (location and number of bands) and compare it to the reference fingerprint (figure 17.1) developed for various bacterial species using BOX PCR. Based on this fingerprint comparison, identify your unknown bacteria species.

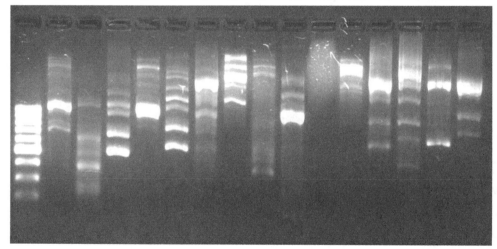

1. DNA ladder
2. *Streptococcus pyogenes*
3. *Pseudomonas aeruginosa*
4. *Proteus mirabilis*
5. *Staphylococcus epidermidis*
6. *Proteus vulgaris*
7. *Klebsiella pneumoniae*
8. *Streptococcus gallolyticus*
9. *Micrococcus luteus*
10. *Mycobacterium smegmatis*
11. *Enterobacter aerogenes*
12. *Staphylococcus aureus*
13. *Escherichia coli*
14. *Citrobacter freundii*
15. *Bacillus megaterium*
16. *Bacillus subtilis*

Figure 17.1. DNA fingerprint profile for various bacterial species.

LABORATORY REPORT

1. What are the three steps of a typical PCR reaction called? What is the purpose of heating the DNA in 1st step of PCR reaction cycle? _____

_____

_____

2. What is the function of Taq-polymerase during PCR? Why would a thermostable taq-polymerase isolated from a thermophilic bacterium (eg *Thermus aquaticus*) be required rather than using one isolated from a common mesophilic bacteria like *Escherichia coli*? _____

_____

_____

3. Fill in the following table with the function of each PCR supermix component.

PCR reaction component	Function
Taq-polymerase	
PCR buffer	
Primer	
dNTPs	
MgCl$_2$	

4. Fill in the following table with the appropriate temperature indicated for each PCR reaction step.

PCR cycle	Appropriate temperature
Denaturation	
Annealing	
Extension	

5. After the completion of one PCR cycle, a single double stranded DNA molecule has been copied into two identical double stranded DNA molecules. Suppose you start a PCR cycle with 10 molecules of DNA, how many total copies of DNA would it be possible to have after the completion of 5 PCR cycles?

_____

_____

6. How are DNA fragments separated using agarose gel electrophoresis? _____

_____

_____

7. What is the purpose of adding ethidium bromide to the gel? _____

_____

_____

8. Why was sample-loading buffer added to your samples before loading it into the well in the agarose gel? How you ensure that your DNA was moving through the gel during electrophoresis? _____

_____

_____

9. How can you determine the molecular weight of DNA fragments using agarose gel electrophoresis?

_____

_____

10. Use the gel image below to fill in the corresponding table with the number of bands in each respective lane and estimate the molecular weight (use the DNA ladder to the left side of the image for estimating molecular weight).

Lane	# of band	Molecular weight (s)
1		
2		
3		
4		
5		

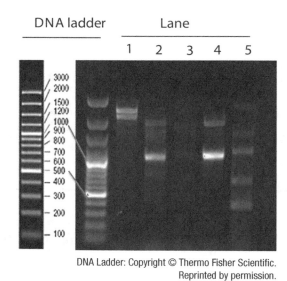

DNA Ladder: Copyright © Thermo Fisher Scientific.
Reprinted by permission.

Exercise 18 Alcohol and Yogurt Fermentation

Background

Fermented foods are some of the oldest and most widely consumed. There are three basic requirements for fermentation: simple sugars, yeast and an anaerobic environment. Microbial fermentations can provide characteristic flavors, aromas, and textures of certain foods, or increase the nutritional value. Fermentation is also utilized as a means of food preservation. The acidic products of microbial fermentations are important in the inhibition of other microorganisms. Fermentation extends the length of time perishable foods, like milk, will remain safe for consumption.

Yeasts like *Saccharomyces* are used to ferment sugars found in fruits and grains into alcoholic beverages. Wine is basically fermented fruit juice. Grapes are a common fruit used for this type of alcohol fermentation but a wide variety of fruits including berries and melons can be used. Under anaerobic conditions, yeasts will convert the simple sugars of fruits into ethyl alcohol, releasing carbon dioxide (CO_2) gas as a by-product. *Saccharomyces* yeasts are able to reduce sulfates in the environment for a source of sulfides for amino acid production. Unfortunately for winemakers, any unassimilated sulfides are converted to Hydrogen sulfide (H2S), which can give wine a "rotten egg" quality. If oxygen is allowed into the fermentation chamber, spoilage bacteria like *Acetobacter* can convert the ethyl alcohol into acetic acid (vinegar).

Saccharomyces yeasts are also used in bread production. The CO_2 produced through their fermentation of the starches in bread dough is what causes the bread to rise. The majority of the ethyl alcohol burns off during cooking, but some residual alcohol remains in the bread.

Lactic acid bacteria (LAB) are involved in the formation of various nonalcoholic products such as yogurt, cheese, sour dough breads, and pickles. Yogurt is a fermented dairy product formed through the metabolic activities of LAB such as *Lactobacillus* and *Streptococcus*. The primary ingredient in yogurt is milk, but the overall composition can be adjusted to achieve the desired content of fat and solids.

The milk mixture is inoculated with starter cultures of bacteria. Heat treatment reduces the number of spoilage organisms by providing a better environment for the starter cultures to grow. The function of the starter cultures is to ferment milk sugar (lactose) into lactic acid. This activity decreases the pH, which denatures milk proteins (casein) and causes the milk to clot. The fermentation of the milk sugars produces the tart flavor that is characteristic of yogurt.

Other bacterial cultures can be added to yogurt as probiotics. These are cultures that may benefit human health by improving gastrointestinal function or stimulating the immune system. The ingestion of these microbes may allow them to colonize and change the composition of the normal flora of the intestinal tract. Probiotic microbes include various species of *Lactobacillus* and *Bifidobacterium*.

Objectives

Upon completion of the exercise students should be able to:

1. Describe the applications of microbes in food production.
2. Demonstrate the effect of alcohol fermentation on fruit juice.
3. Demonstrate the effect of lactic acid fermentation on milk.
4. Utilize basic stain techniques to isolate bacteria from yogurt sample.

Materials

Alcohol Fermentation
> grape juice
> dry yeast
> 250 ml flask
> small balloon

Yogurt Fermentation
> powdered milk
> liquid milk
> yogurt with live cultures
> graduated cylinder
> pH paper
> 400 ml beaker
> microwave
> plastic wrap

Methods

1st Laboratory Session

A. Alcohol Fermentation

1. Observe and record the aroma and pH of the grape juice before your instructor sets up the alcohol fermentation demonstration.

2. After the grape juice has been exposed to the yeast, sealed, and incubated, observe and record the aroma and pH and test for hydrogen sulfide (H_2S).

B. Yogurt Fermentation

1. Measure 100 ml of liquid milk into a beaker. Record the pH and information about color, texture, and aroma of the milk on your laboratory report.

2. Add 4 grams of powdered milk to the liquid milk.

3. Place the milk mixture into the microwave. Heat the milk mixture using 15 second cycles until the powdered milk is completely dissolved. Watch the milk mixture carefully. Once bubbles begin to form, it may boil over very quickly.

4. Remove the milk mixture from the microwave and allow it to cool until you can handle the beaker comfortably with your bare hand.

5. Add 1 teaspoon of yogurt to the milk mixture to introduce fermenting bacteria and stir the mixture well. Cover beaker with plastic wrap and incubate at 37°C.

2nd Laboratory Session

6. Examine your yogurt culture from the last laboratory session.

7. Record information about pH, texture, color, and aroma on your laboratory report.

8. Prepare a smear and perform a simple stain with methylene blue. Refer back to exercise 5 to review the simple stain procedure.

9. Record the microscopic appearance of any organisms detected.

LABORATORY REPORT

Alcohol Fermentation

1. Record the data for both the unfermented and fermented grape juice in the table below.

	pH	Aroma	H$_2$S
unfermented juice			
fermented juice			

Questions

2. What component of the juice is being fermented and what are the end products of this fermentation?
_____
_____

3. Why should you be concerned with H$_2$S production during wine production?
_____
_____

4. If wine is spoiled by Acetobacter, what would the end result be? How would this affect the taste of the wine? _____
_____
_____

5. If yeasts are involved in fermenting bread dough, why do you not get drunk from eating a sandwich? Is there any alcohol at all in breads? _____
_____
_____

Yogurt Fermentation

6. Record the data for the unfermented milk and the yogurt product in the table below.

	pH	Color	Aroma	Texture
Milk				
Yogurt				

7. Draw and describe the microscopic appearance of any organisms observed with the simple stain.

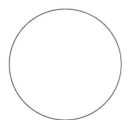

Morphology

_____

_____

Questions

8. What type of microbial fermentation is used in the production of yogurt?

_____

9. Why might microbes not involved in the fermentation processes be added to the yogurt? _____

_____

_____

10. How might the fermentation of foods by microbes be useful in food preservation?

_____

_____

Exercise 19 Root Beer Production

Background

The production of root beer dates back to the American colonies and employs the same type of alcohol fermentation utilized in the production of beer and wine. *Saccharomyces cerevisiae* yeasts produce ethyl alcohol and carbon dioxide waste products through the process of fermenting sugar. The same carbon dioxide that forms the bubbles in beer and causes bread to rise supplies the carbonation of root beer. The characteristic flavor of root beer traditionally was achieved through the addition of sassafras root, but as it was discovered to be a carcinogenic, cancer-inducing compound, we will use root beer extract instead. The substitution of vanilla extract will result in the production of cream soda. In this exercise, we will use a 2 liter bottle to create a closed fermentation chamber in order to fashion a fermented root beer or cream soda beverage.

Objectives

Upon completion of the exercise students should be able to:

1. Demonstrate the applications of microbes in food production.
2. Utilize fermentation properties of yeast to produce root beer and cream soda.

Materials

    clean, empty, 2 liter plastic bottles with caps
    funnel
    distilled water
    bakers yeast
    root beer extract
    vanilla extract
    sugar
    small graduated cylinder
    top loading balance
    weigh dishes

Methods

1. In beaker, dissolve 1.2 grams of yeast in 50 ml warm water. If the water is too hot, the yeast may be killed. Let stand for 5 minutes.
2. Using funnel add 230 grams of sugar to the 2 liter bottle.
3. Using funnel again, add 7 ml of root beer extract or vanilla to the sugar.
4. Add the yeast to the mixture in the 2 liter bottle, then add enough cool water to finish filling the bottle (leave about 2 inches from the top). Rinse the yeast and extract from the funnel as you fill the bottle. Place the lid tightly on the bottle.

5. Incubate the root beer or cream soda at room temperature. <u>Bottles should be checked daily for tightness, if they become too pressurized, they may explode</u>. When the bottle first begins to feel tight (pressurized) (typically around day 4) place it in the refrigerator until your next lab period. (The cold temperature stops the fermentation.)

6. Do not shake the bottles and open them carefully releasing the pressure a small bit at a time. Taste the beverage and record your observations. (Caution: there will be trace amounts of alcohol in these beverages. It would require that an individual drink over a gallon to equal the alcohol content of one 12 oz beer, but this amount could still be dangerous to individuals with an allergy to alcohol.)

Adapted from Dr. David Fankhauser, University of Cincinnati posted on http://biology.clc.uc.edu/fankhauser/Cheese/ROOTBEER_Jn0.htm

LABORATORY REPORT

1. What process is yeast using to make energy in this experiment?

   _____

2. Why was the sugar necessary in this experiment? _____

   _____

   _____

3. Why was it necessary to add warm water to the yeast before you added it to the sugar solution? _____

   _____

   _____

4. Why were you told to be cautious about how hot the water was which was used to suspend the yeast?

   _____

   _____

5. Is there any alcohol in the beverage you produced? _____

   _____

6. If the same process is used to make bread, is there likely to be very much alcohol in bread? Why or why not?

   _____

   _____

7. Why do we want the carbon dioxide to collect in the bottle? _____

   _____

   _____

8. What effect does carbon dioxide produced during fermentation have on the production of bread?

   _____

   _____

9. Why did putting the bottle in the refrigerator stop the fermentation process?

   _____

   _____

10. Record observations of taste, appearance, smell and level of carbination for your beverage.

    _____

    _____

    _____

20 Epidemiology–Synthetic Epidemic

Background

Epidemiology is a science that studies many aspects of diseases within populations. Epidemiological studies include those directed at determining the causative agent and reservoir of infectious disease (those caused by pathogens) as well as identifying the method of transmission, incidence, and distribution of a disease within a population. Studies are also directed at detecting patterns associated with outbreaks of disease and establishing methods for controlling future outbreaks. The term *epidemic* indicates an increase in the incidence of disease in a particular population during a specific period of time. Epidemics are classified as common source or propagated. Common source epidemics result from a group of people being exposed to a pathogen simultaneously, such as in contaminated food or water, and is characterized by a sudden increase in the incidence of disease. A propagated epidemic occurs when a pathogen is transmitted from one host to another and results in a steady increase in incidence of disease over time. The first case in a propagated epidemic is considered the index case. Infectious diseases that can be transmitted between susceptible hosts are termed communicable. Modes of transmission for communicable diseases include direct contact and indirect contact through inanimate objects called fomites such as tissues or doorknobs. In addition, some pathogens are spread through droplet transmission or through a nonliving vehicle like contaminated water or food. Several insects, especially arthropods like mosquitoes and ticks, serve as vectors of disease transmission. In this exercise you will generate a synthetic epidemic and then carry out a basic epidemiological study to determine the index case and incidence of disease in your population.

Objectives

Upon completion of the exercise students should be able to:

1. Explain the basic principles of epidemiology.
2. Differentiate between the various modes of transmission for infectious disease.
3. Describe what is meant by the terms propagated epidemic and index case.
4. Identify patterns in disease transmission in order to determine the source of infection.

Materials

    numbered test tube containing baby powder or corn starch
    iodine spray
    sterile starch-free filter paper disk

Methods

1. Select a numbered test tube.
2. Draw a line down the center of the filter paper disk and label one side with the number 1 and the other with the number 2.

3. Moisten the palm of your dominant hand with water, then coat it with the powder in your numbered tube. Be careful not to touch any surface with the powder-coated hand.

4. When called on by your instructor, shake hands with a classmate. Be careful not to use excessive motions, but shake as you normally would. It is best not to choose your immediate neighbor. You will only be called on to shake hands once, but you may be selected by others to shake multiple times.

5. After each person has shaken hands at least once, wipe your hand across the filter paper on the side labeled with the number 1. Be careful not to touch the other side of the paper.

6. Repeat the process through a second round of hand shaking, being careful not to touch any additional surfaces with the powdered hand. Do not add more powder between hand-shaking rounds.

7. Once the second round of shaking is complete, wipe your hand across the side of the paper labeled with the number 2.

8. Lightly spray the filter paper with the iodine solution. If the paper turns bluish-black, you have been exposed to the pathogen (starch). If the paper is brown or yellow, you have not been exposed.

9. Record the entire class's results for each round of shaking, then attempt to determine who the index case was. This can be done by identifying all positive cases and working backward. A pattern should develop which will lead back to the source of the "infection."

LABORATORY REPORT

1. Record your results for each round of hand shaking in the table below. The shaker is the person who is instructed to shake hands while the shakee is the person they chose to shake hands with. Identify the shaker and shakee by using the number on their test tube. Results for each round of shaking should be recorded as positive (+) or negative (−) for each individual.

Shaker Round 1	Positive or negative	Shakee Round 1	Positive or negative	Shaker Round 2	Positive or negative	Shakee Round 2	Positive or negative

2. Who in your lab section was the index case (source of epidemic)?

_____

3. How many students became carriers after the first round of hand shaking?

_____

4. How many carriers were there in your "population" after the second round?

_____

5. What is the definition of the term epidemic?

_____

_____

6. What is the difference between a propagated and a common source epidemic?

_____

_____

7. Would your synthetic epidemic be considered common source or propagated?

_____

8. What was the mode of transmission for the "pathogen" in this epidemic?

_____

9. What would have been effective in stopping the transmission of this "pathogen" within your "population"? _____

10. What are other possible modes of transmission for infectious disease? _____

_____

_____

References

Brown, A. (2007). *Benson's Microbiological Applications: Laboratory Manual in General Microbiology* (9th ed.). New York, NY: McGraw-Hill.

Carrington, E. (2009) *Microbiology Laboratory Manual: For Students Entering into the Allied Health Professions* (2nd ed.). Dubuque, IA: Kendall Hunt.

Pollack, R., Findlay, L., Mondschein, W., & Modesto, R. (2009). *Laboratory Exercises in Microbiology* (3rd ed.). Hoboken, NJ: John Wiley & Sons.

Tortora, G., Funke, R., and Case, C. (2010). *Microbiology: An Introduction* (10th ed.). San Francisco, CA: Benjamin Cummings.

Winfrey, M., Rott, M., Anglehart, S., Schwan, W., Taylor, B., Burns, R. & Grimes, J. (2004). *Laboratory Manual for Fundamentals of Microbiology*. University of Wisconsin-La Crosse.

Figure 3.2. Various bacterial and fungal colonies (third row is a top and bottom view of a plate with multiple fungal colonies).

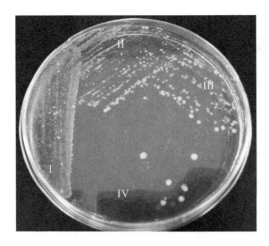

Figure 4.2. Isolated colonies on a typical streak plate.

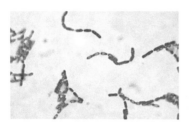 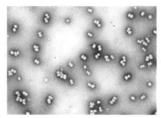 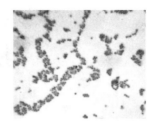

Figure 5.5. Typical simple and negative stain reactions.

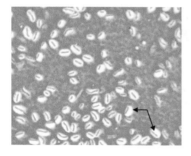

Figure 6.1. Capsules appear as clear areas around bacilli.

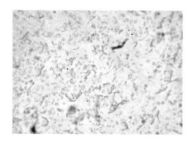

Figure 6.2. A typical endospore stain showing multiple free endospores and vegetative bacilli.

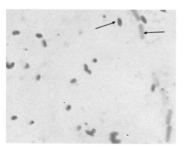

Figure 6.3. The green structures represent endospores and the red structures are bacillus shaped vegetative cells.

Reagents	Time Applied	Reactions	Appearance
Unstained smear			Cells are colorless and difficult to see.
Crystal violet	1 minute, then rinse with water	Basic dye attaches to negatively charged groups in the cell wall, membrane, and cytoplasm.	Both gram-negative and gram-positive cells are deep violet.
Gram's Iodine (mordant)	1 minute, then rinse with water	Iodine strengthens the attachment of crystal violet to the negatively charged groups.	Both gram-negative and gram-positive cells remain deep violet.
Alcohol or acetone-alcohol mix (decolorizer)	10 to 15 seconds, then rinse with water	Decolorizer leaches the crystal violet and iodine from the cells. The color diffuses out of gram-positive cells more slowly than out of gram-negative cells because of the chemical composition and thickness of the gram-positive cell walls.	Gram-positive cells remain deep violet, but gram-negative cells become colorless and difficult to see.
Safranin (counterstain)	1 minute; then rinse thoroughly, blot dry, and observe under oil immersion	Basic dye attaches to negatively charged groups in both cell types. Few negative groups are free of crystal violet in gram-positive cell, whereas most negative groups are free in gram-negative bacteria. Consequently, gram-positive bacteria remain deep violet, whereas gram-negative bacteria become pink or red.	Gram-positive cells remain deep violet, whereas gram-negative cells are stained pink or red.

Figure 7.1. Gram Stain Procedure.

© Kendall Hunt Publishing Company

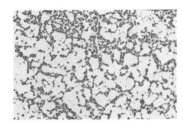

Figure 7.2. *Staphylococcus aureus* gram reaction.
From *Microbiology*, 3/E by Daniel Lim. Copyright © 2003 by Kendall Hunt Publishing Company. Reprinted by permission.

Figure 7.3. *Escherichia coli* gram reaction.
Source: Centers for Disease Control and Prevention

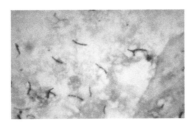

Figure 7.4. Acid fast bacilli.
Source: Centers for Disease Control and Prevention

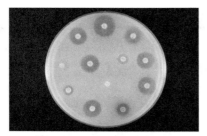

Figure 12.1. Multiple anti-microbial saturated disks producing various size zones of inhibition against a lawn of bacterial growth.
Source: Centers for Disease Control and Prevention

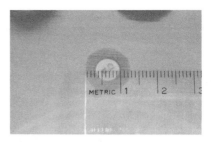

Figure 12.2. Each zones of inhibition is measured to determine the level of sensitivity an organism displays to a given anti-microbial.
Source: Centers for Disease Control and Prevention

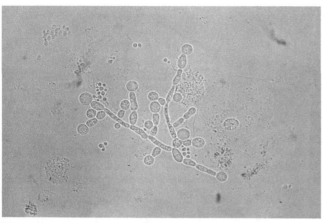

Figure 13.4. *Candidia* yeast form pseudohyphae when buds fail to separate.
Source: Centers for Disease Control and Prevention

Figure 13.5. Conidiaspores forming on phialides formed by the blue-green *Penicillium* mold.
Source: Centers for Disease Control and Prevention

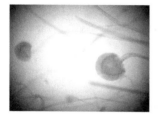

Figure 13.6. Asexual sporangiospores produced by *Rhizopus*, a common bread mold.

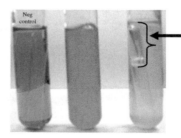

Figure 15.1. Tube on left is negative control. Middle tube (lighter pink than control) is negative for acid production. Tube on left is positive for both acid and gas production.

Figure 15.2. a. MR test: Tube on left is negative control. Middle tube shows a negative reaction and tube on the right is positive. **b.** VP test: Tube on right is negative control. Middle tube shows a negative reaction and tube on the left is positive.

Figure 15.3. Tube on left is negative control. Middle tube is positive for citrate utilization. Tube on right reveals a negative result.

Figure 15.4. Tube on left is negative for nitrate reduction as demonstrated by the bight red color after the addition of both nitrate A and B and zinc powder. The dark red tube on right is positive for the reduction of nitrate to nitrite.

Figure 15.5. Tube on left is negative for nitrate reduction as demonstrated by the bight red color after the addition of both nitrate A and B and zinc powder. The colorless tube on the right is positive for the reduction of nitrate beyond nitrite as no color developed with the addition of zinc.

Figure 15.6. Tube on left is positive for oxidase as demonstrated by the purple color. The tube on right is negative.

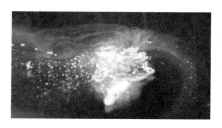 

Figure 15.7. Colony on left is positive for catalase as demonstrated by bubble production. The colony on right is negative.

Figure 15.8. Bacteria on the right display starch hydrolysis (positive for amylase) while the bacteria on the left are negative for amylase production.

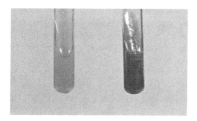

Figure 15.9. Tube on the right is positive for urease production while the tube on the left is negative.
Source: Centers for Disease Control and Prevention

Figure 15.10. Tube on left is positive for phenylalanine deaminase. The middle tube represents a negative control. Tube on right reveals a negative result.

Figure 15.11. Tube on left is positive for indole production. Tube on right reveals a negative result.

Figure 15.12. Tube on left is positive for indole production and motility. Middle tube is positive for indole but reveals no motility. Tube on right is negative for indole but is positive for hydrogen sulfide production and motility.

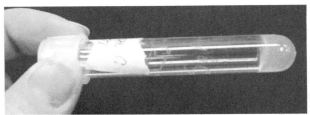
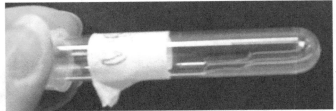

Figure 16.4. The tube on the left revels a positive reaction (clotting of rabbit plasma). This indicates that the organism produces coagulase. The tube on the right revels no clotting which is a negative result.

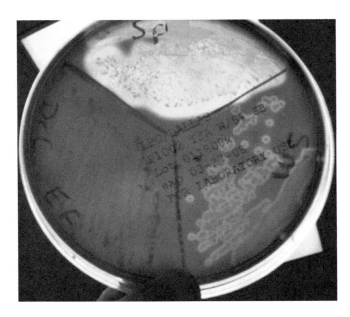

Figure 16.5. Hemolysis of blood agar. The top quadrant depicts complete (beta) hemolysis. The quadrant on the right shows non-hemolytic (gamma) colonies while the left quadrant shows the brownish-green deposits associated with incomplete (alpha) hemolysis.